Heinz Jenuwein

Avocado, Banana, Coffee

How to grow useful exotic
plants for fun

82 colour illustrations

Adapted and translated by
Jean and Edmund Launert

British Museum (Natural History)

Cover: Litchi with mature fruits (see p. 79).
Photograph on p. 2: A conservatory ideal for the cultivation of tropical and subtropical plants.

© 1986 Eugen Ulmer GmbH & Co.
Wollgrasweg 41, 7000 Stuttgart 70 (Hohenheim)
© 1988 English Edition British Museum (Natural History)
First Published in England 1988
British Museum (Natural History)
Cromwell Road, London SW7 5BD

British Library Cataloguing in Publication Data

Jeunwein, Heinz
    Avocado, banana, coffee.
    1. Great Britain. Exotic plants.
    Cultivation
    I. Title   II. British Museum (Natural History)
    III. Avocado bis Zuckerrohr.
    English
    635.9'52

ISBN 0-565-01040-9

Printed in Germany

Contents

Preface

This book attempts to fill a gap still left in the impressive mass of horticultural literature. Although there are several good descriptive works on the subject of useful exotic plants, the amateur grower will not find instructions as to how to cultivate them. The hobby of growing exotic plants of this kind may start when, through curiosity, we pop a few orange or lemon pips into a flowerpot. Perhaps it will be an avocado stone that starts us off, but we probably fail to nurse the ensuing plant into maturity. With so many exotic fruits now on the market thanks to modern air transport, it is not surprising that the consumer wants to see what the mother plant looks like and will therefore attempt to grow one in our latitudes.

Cultivation instructions in professional literature, if at all available, generally refer to conditions in the original habitat. This is why they differ markedly from those given in this book. Our author has confined his selection of useful tropical plants to species he has grown in Europe over a number of years and his advice relates to our conditions. His choice is not restricted to edible plants but gives some species useful to man in other ways, e.g. fibre producing plants. Some plants can be accommodated in the living room, but for the majority a heated greenhouse will be necessary.

Written originally for a Central European readership, the book required a few alterations to suit the climatic conditions of the British Isles. Moreover, the author frequently encourages his readers to import seeds or living material from the country of origin. We must point out, however, that there are strict customs regulations regarding such imports into Britain, and the traveller is therefore strongly advised to consult the authorities beforehand at the following address.

Plant Health Division,
Ministry of Agriculture, Fisheries and Food,
Great Westminster House,
Horseferry Road,
London SW1P 2AE.
(Tel. 01-216 6174)

If the enthusiast wishes to obtain plant material from a botanic garden, as also encouraged by the author, he should obtain prior permission rather than be caught loitering with intent armed with knife and scissors!

The potential grower of useful exotic plants will meet with many difficulties but when he succeeds, the results will be rewarding.

Jean and Edmund Launert

Introduction

Most people at some time in their lives long for a garden. Since this dream cannot always be realized, many turn to the cultivation of indoor plants, for example, the ubiquitous rubber plant and cyclamen. The first plant I grew was coffee, but no matter which you choose the first step is the hardest.

Special precautions

To prevent anyone from throwing his own spanner in the works, I would like to bring to the reader's attention certain mistakes that I myself have made, indeed had to make, since there is surprisingly little literature on the subject of the

The loquat (centre) will only grow well in a greenhouse.

Plants of the
tropics and
subtropics are
hardier than one
usually imagines:
here *Musa basjoo*
amidst olives.

cultivation of useful tropical or subtropical plants. Any information as to a plant's requirements is limited to a description of the habitat in which it is cultivated in the foreign country. Nowhere is there any mention of the possibility and means of growing this same plant under conditions provided by the enthusiast in our climate, nor of the success rate. There is nothing more depressing than to acquire an exotic plant at great expense and then to watch it die off bit by bit every day.

A common mistake is to take as gospel exact and rigid instructions for the care of this or that plant. The plant lover who sticks unimaginatively to such directions will often experience disaster. The living plant is not an automaton which performs in this or that way according to what is fed into it. For the successful care of problem plants – almost exclusively cultivated plants of the tropical and subtropical regions – the main thing is to possess a certain sensitivity to their needs. Given patience and time they will reward the cutivator with successful growth.

Watering as an example

Watering may be a simple act, but it plays a vital role in plant care.

Tap water contains dissolved mineral salts in various quantities (calcium and magnesium salts) i.e. it is more or less "hard", and therefore its suitability for the continuous watering of pot plants will vary according to the case. The plant's constant need for water will lead to calcium deposits in the potting material which will have consequences on the plant's health.

Every plant is dependent on the chlorophyll in its leaves. This chlorophyll can only be manufactured with the help of the iron present in the soil. Absorbable iron must therefore be present in the soil for the plant to use. Only the tiniest quantities are necessary, hence the term "trace elements". The iron itself is not present in the chlorophyll but functions as a catalyst, i.e. it influences and guides its formation. In potting material with calcium deposits and a resulting low pH value, the iron is no longer available to the plant, but bound in the material. The consequence is that no more chlorophyll is manufactured, the leaves become light green with clearly visible veins which are still dark green, then yellowish, and in severe cases whitish. Finally the plant loses all its leaves and dies. Some gardeners think that loss of leaves means a lack of water, increase the quantity and the process of dying is merely hastened.

Basic rules for plant care

If one is interested in cultivating unusual plants it is necessary to familiarize oneself with a few basic rules for plant care. There are no hard and fast rules when dealing with plants. Accumulation of personal knowledge, along with keen observation, is the only formula for success.

The water

The water we give our plants should have as low a lime content as possible; in horticultural terms it should be "acid".

pH-value

The acidity of the water is measured in pH-values. The pH-value scale ranges from 1 to 14. The 7 mark is referred to as the neutral point; over 7 and the water is alkaline, under 7 and it is acid. Water with values between 7 and 14 is useless for our purposes. A pH-value of 1 represents an acidity grade equalling that of vinegar, of 4 that of a good wine. Water with a pH value of 7 is unsuitable for our task. Rain-water has a pH-value of 5.5 to 6.5 and is ideal. Water with pH values of 4 to 5.5 is most suitable for swamp plants, especially the interesting carnivorous ones.

Water in the range of 5–7 is therefore most generally suited to

our purposes. Rain-water is ideal, if it can be collected outside large towns. In cities the air is often so polluted that rainwater eventually accumulates damaging contaminants.

Measuring the pH-value

We can easily find out for ourselves whether the water possesses a suitable pH-value for our plants. Any chemist will sell us litmus paper complete with colour key. Detach a test strip, dip it into the water and after the few seconds it takes for the chemical reaction to occur, compare it with the colour key. Each colour grade on the key will be given its corresponding pH-value, so it is simple to ascertain that of our sample.

Water softeners

When rain-water is in short supply softened tap water may be used. Garden centres will provide water softeners, which when used in accordance with the manufacturer's instructions will render the water suitable for our needs. Phosphoric acid or nitric acid obtainable from chemists and added in drops to the water will also do the trick. Phosphoric acid encourages root and flower development, and nitric acid will, through chemical changes in the soil, be converted into nitrogen which can then be used by the plant.

The water given to the plants, particularly to delicate plants, should be at room temperature.

Potting materials

The plants which we cultivate grow, with very few exceptions, in aerated soil rich in humus. Even plants which grow on stony bare ground in their natural habitat benefit when the soil is rich in nut-

rients. At the same time care must be taken that the latter do not become too luxuriant and lose their characteristic habit.

With citrus species, which in their usual areas of cultivation normally grow in heavy loamy soil, a loose material containing humus is necessary to produce a fruit with the desired peel thickness. In general with all citrus species the lighter the soil, the thinner the fruit peel (see also the heading 'Soil' in the individual plant descriptions from p. 15 onwards).

As an experiment, young plants of the most varied tropical and subtropical species were planted in plain peat. Light feeding provided the nutrients lacking in the peat and the plants flourished marvellously.

To test the pH of the potting material we can also use the litmus test. The strip must be pushed into the soil with a wooden or metal stick and not the finger so that a true reading is obtained.

A widely spread belief must be noted here. If the surface of the potting material becomes covered with a blackish-green layer, gets slimy and growth seems impeded, it is generally assumed that the material has 'gone sour' and has become acid. The opposite is true. The soil mixture has become alkaline because of the lime in the water that has been applied. Only immediate repotting in fresh soil will save the day.

Where both space
and means permit
a greenhouse is
ideal for exotic

plants, offering
them optimal
light in all
seasons.

Light

The light requirements of species vary so considerably that this subject is dealt with from p. 15 onwards for each individual plant. Exceptions to the general rules given in horticultural literature will be noted. The reason for this is that light measurements for the native land are often given, whereas here account must be taken of the requirements the tropical and subtropical plants have in our latitudes.

Example: the Coffee Plant

The coffee plant may be taken to demonstrate this fact. In the literature, shady growing conditions are recommended for coffee; only in very cloudy regions may the need for shade be ignored. Indeed, too strong a light will, in the case of coffee, lead to overproduction and premature exhaustion, an important consideration for the plantation owners.

The European plant enthusiast will also want to see his coffee plant flower and fruit. It will do so, contrary to the general advice, when it gets lots of sun, assuming that it is mature enough. It is clear that the power of the sun in the tropics cannot be compared with its intensity in our climes. We may therefore place our coffee plant in full sunlight without reservation,

and in the height of summer it may stand outdoors.

Full light in autumn and winter

From September we can expose our plants to maximum light, also those under glass, for the shoots must mature. Supplementary light in the winter months is only necessary for particularly delicate plants and seedlings. The trade can supply all that is necessary in this respect.

In general the plants must not be allowed to grow in winter. This would result in weak, thin shoots which cannot support themselves and have to be pruned off in spring. Purely tropical species which are adapted to growth all year round cease this activity automatically when the light is reduced and the temperature falls.

11

After overwintering

Care must always be exercised when bringing plants out into the sun after overwintering. Only plants which come through winter without leaves can stand full light, all others suffering from burns.

Containers

In defence of the plastic pot

Whether to use a terracotta or a plastic pot is a controversial point. Both have their advantages and disadvantages. Personally I have come down in favour of the plastic version. It is lighter, a point to consider when we think of the load the greenhouse shelves have to bear. Moreover, plastic containers need less attention and are therefore labour saving.

The nonporous wall of the plastic pot inhibits evaporation. Less watering is therefore required and the soil ball remains evenly moist, a benefit to growth. At the same time, care must be taken when watering to note whether the soil at the bottom of the pot is adequately moist, even if the surface is already dried out. The often mentioned finger tip test is no good here: it will show us that the top few centimetres are dry, and we water. Since the middle still has plenty of mois-

ture our watering only results in stagnant saturation at the bottom and the roots will begin to rot. A good idea is simply to pick up the potted plant and its weight will tell us whether the soil really needs water.

A further reason for choosing the plastic pot is that lateral evaporation from the pot does not take place as it does in the terracotta version, so there is no subsequent cooling of the soil ball.

Temperature

Overwintering

As the information on the individual plants from p. 15 onwards will show, most of them, with a few exceptions, can survive the winter at amazingly low temperatures. They are not fed, of course, during this period and receive only sufficient water to prevent the soil ball from drying out.

Too high a temperature in combination with poor light during our winter months only leads to weak growth. In *Citrus* species, with the exception of the pummelo and lime which are adapted to tropical temperatures, overwintering just above freezing point is indeed an advantage, since low temperatures and short days encourage flower development.

Day and night temperatures

In general, night temperatures are a few degrees cooler than the current daytime temperatures. In the summer months temperatures in the greenhouse exposed to the sun's rays can reach 30 to 35°C, but are easily tolerated. Shade must be provided, however, to prevent further rises. Temperatures over 40°C lead to the scorching of soft plant parts.

Supplying nutrients

In many manuals advice is limited to "feed", but nowhere are quantities and type of feeds given. This is for the following reasons:

When long-lived plants are grown in pots, chemical feeding eventually leads to an increasing mineralization of the soil. In the process of manufacture, chemical nutrients are coupled with carrier substances which accumulate in the soil since they are not used by the plant.

Manure as a feed

This, especially if it can be obtained from a zoo, is an excellent source of nutrition for the plants and does not carry the same risks as chemical feeding. Manure obtained from intensive rearing units and broiler houses is to be avoided because it has antibiotic residues which will kill off all micro-organisms in the potting material, hence the recommendation for the zoo.

Zoo manure will contain nothing harmful to our plants, and most zoos will supply it, if requested. For simple feeding cattle dung is good, for example, that of the water buffalo. To make humus, elephant dung is ideal. It has fewer ingredients but many digested particles and when strewn on the floor of the greenhouse it will form an excellent fertile humus layer in a short time.

Feeding

A mixture of a small amount of manure (for example, cattle dung) in rain-water can be liberally applied as a feed to the plants, and will result in strong, healthy growth. The mixture should be left to stand for a day before use. This is a practice which was common in agriculture before the chemical craze took hold and it still shows the best results. I have harvested from 15 to 25 kg of lemons from one tree, but only since I gave up chemical feeding.

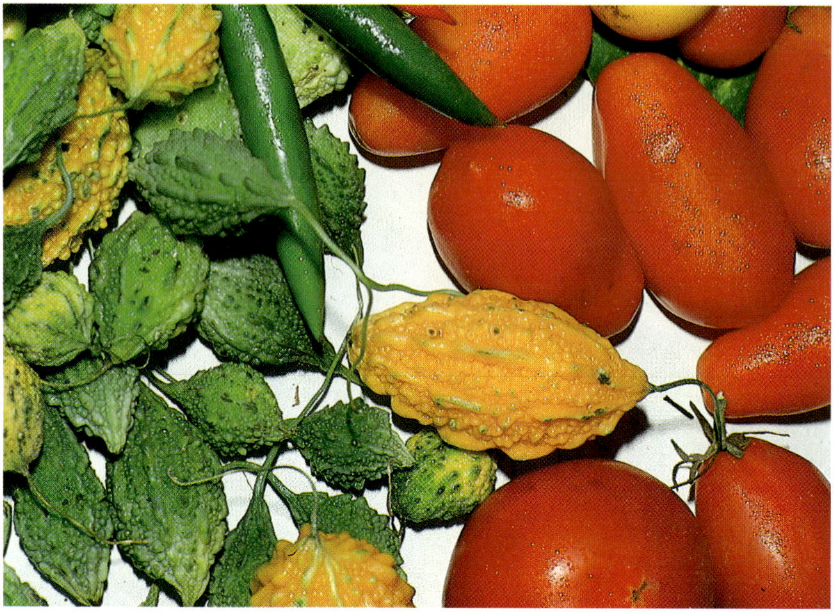

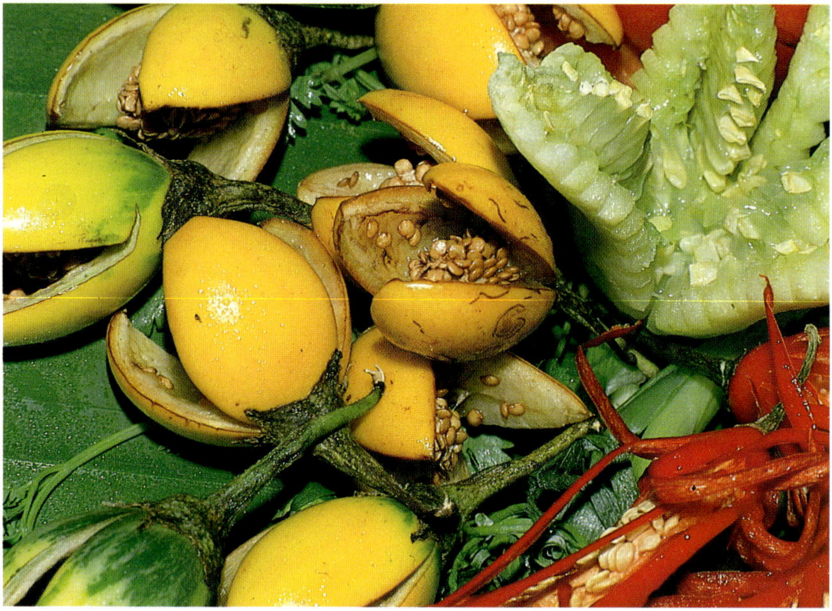

Plant manual

The following plants described individually are introduced by their English names, but the order in which they appear depends on their botanical names.

Pineapple guava or feijoa
Acca sellowiana (syn. *Feijoa sellowiana*)
Family: Myrtaceae

Origin: South America. The plant has subsequently been introduced into the subtropical regions of Asia and Africa.

Habit: The plant grows in our climate as a 2 m high, many-branched, dense shrub. The small elliptic leaves are dark green on the upper side and grey-green underneath. The decorative flowers which are formed in the leaf axils are dark red with long protruding stamens. They can be self-pollinating. Small, sourish but very aromatic berries develop and these may be eaten raw.

Left: Provided the plants are properly cared for, tasty fruits can be produced even in temperate climates.

Right: The South American feijoa produces small, rather acidulous but aromatic berries which are eaten raw.

Habitat: As the thick, water retentive leaves indicate, the plant is adapted to a hot, dry habitat. For this reason we give it the sunniest possible position. In wet summers it must be sheltered from the rain, otherwise the roots will be damaged. It thrives well under glass but over-generous amounts of water along with a high degree of warmth will destroy the plant's characteristic appearance. It is best to stand it in the open air in summer. With the onset of autumn rain it should be moved to a drier place but given as much light as possible. In winter it needs a light, cool position.

Soil: The feijoa will do best if the soil is not too rich. Leaf mould mixed with quartz and a slight addition of peat is the correct mixture. It is important to see that superfluous water drains off.

Watering: Water with restraint; let the soil ball almost dry out then water thoroughly. Water allowed to stand in the pot will kill the plant. If too much water is given the plant dies; too little and it loses its leaves. Since the feijoa is prolific in its leaves there will be a high water loss even in winter, hence only experience and sensitivity to its needs on our part will ensure that the plant survives the winter.

Feeding: Only light feeding is required. One manure feed every four weeks during the growth period is sufficient. Cease feeding completely at the end of August to allow the shoots to mature.

Maturing, harvesting: The berries are ready for eating in autumn, when they will have become soft. They are best eaten raw.

Propagation, cultivation: If it has not been possible to obtain seeds from the country of origin it may be possible to obtain some berries from a botanical garden. These are stored in warm, dry conditions through the winter, immersed in warm water for a few days in spring and then sown in sandy soil about 1 cm deep. They will germinate at a temperature of about 20 to 25°C after 3 or 4 weeks. Occasionally young plants are offered for sale in gardening catalogues in spring.

Kiwi, Chinese gooseberry
Actinidia chinensis
Family: Actinidiaceae

Actinidia, commonly known as the kiwi, is usually sold in garden centres. Since a plant is either male or female, one of each sex must be planted if fruits are to be obtained. The seeds inside fruits which are found in our supermarkets all year round can also be used for propagation.

Origin: The kiwi's country of origin is China. The large cultivations which supply the fruit market are, however, mainly in New Zealand.

Habit: The kiwi is a climber with large, soft, heart-shaped leaves which are slightly hairy. The new

Chinese gooseberries, also known as kiwis, are nowadays commonplace and available in abundance.

growth is red, the flowers, arranged in leaf axils, are yellowish and resemble the dog rose. The male flowers have numerous stamens with yellow or purple anthers. The plant is fairly large, attaining several metres in height and width. The fruit of the cultivar 'Hayward' most usually found in the shops is roughly 8 cm long and 5 cm wide. Its flesh is bottle green with many small seeds which are eaten with it. The plants are ready to flower after 3 years, the fruits are harvested in November and will stay fresh in the refrigerator for several months. The long branches of the kiwi plant do not cling by themselves but must be attached to a support.

Habitat: Since the kiwi plant is too large for the average garden greenhouse, only an open air position can be considered. In spite of the claims of some horticultural sup-

pliers I would advise against open air cultivation except in a very sheltered position. It can survive for several years but a severe winter will destroy it. However, in warmer parts of England it can prove to be quite hardy.

Cultivation in a tub is essential. In late May the kiwi should be placed in the warmest sunniest place available. It should stay there until the onset of night frosts. If it is caught by the first frost and the foliage has suffered, it does not matter too much, and the foliage will simply be shed. In the winter put the plant in a dry place; in the cellar if available. A cool temperature is all right. A garage which can be kept frost free is also suitable, and it may be dark since the plant is bare. Kept under glass the kiwi acts as a magnet to red spider mite, so be sure to spray in time.

Soil: *Actinidia chinensis* needs a well aerated humus soil. The commercial mixtures are particularly suitable.

Watering: The kiwi needs plenty of water in the summer when standing in full sunlight. The water should be lime-free or at least low in lime content, that is, soft. Rainwater is most suitable. Water collected in cities will, however, contain too many damaging additives, so in this case tap water treated with a softener available in the trade should be given. The soil ball should be kept slightly moist in winter to prevent the roots from dying.

Feeding: During its most active growth period the kiwi may be encouraged with a weekly feed. Since it is not particularly sensitive to salt, it may be fed with a chemical fertilizer in a concentration of 2 g per litre of water.

Maturing, harvesting: When the plants are kept under glass pollination is carried out using a paint brush. After pollination the fruits start to form, and enlarge covered by a brown, furry skin. They are ready in late autumn, and should be harvested before the first frost. At this point they are still hard and may be kept for some months in the vegetable drawer of the refrigerator. They should be brought out into room temperature a few days before eating.

Propagation, cultivation: If you wish to grow plants from the seeds contained in the freshly bought fruit, place the seeds with some of the fruit flesh still attached in warm water for a few days. When the flesh has disintegrated, dry out the seeds. Fill a pot with a mixture of commercial potting material and peat and dampen slightly. Sprinkle the seeds on top, press down gently but do not cover with soil.

Watering will wrench the tiny seeds from their germination bed, so it is better to stretch polythene over the pot to hinder evaporation and retain moisture. Depending on the degree of warmth and light conditions the seeds will germinate after 8 to 14 days. They remain in the pot until the seedlings are about 2 cm high then the first thinning takes place; this is necessary as the seeds are so small that more seedlings will grow than are required. As soon as the seedlings have unfolded their cotyledons the polythene may be removed.

Sisalagave
Agave sisalana
Family: Agavaceae

Origin: Central America, Mexico. These plants are hardly, if ever, commercially available.

Habit: The sisalagave is an evergreen perennial found in the tropics; definitely a dry habitat plant. It has fleshy leaves which form a

Agave sisalana, seen here in its natural habitat, will never flower like this under our conditions.

Soil: It is happy with a relatively poor soil. Commercial potting material mixed with an equal proportion of sand is best. It will not stand much peat in the mixture. Water accumulation in a soil too rich in humus will lead to root rot. It is a good idea to put several large shards at the bottom of the pot to ensure that superfluous water drains quickly.

Watering: Since it is a water-storing plant the sisalagave needs to be watered only when the soil is totally dry. Constant moisture round the roots will cause them to die. In winter it may be allowed to stand completely dry.

Feeding: One feed half way through the growth period is sufficient. The plant should not be allowed to grow too quickly otherwise it will become too large for its winter position.

Maturing, harvesting: *Agave sisalana* is cultivated for is leaf fibres. Machinery is used to extract them. They are 1 to 2 m long and a brilliant yellow colour. They are used in the manufacture of course yarn, string, ropes, mats, furnishing fabrics and carpets. The fibre yielding agaves cultivated in Asia belong to another species.

Propagation, cultivation: This is only possible by means of root shoots. Only the botanical garden or a journey to Mexico can provide these, and, being succulent, they are easy to carry in one's luggage (custom regulations permitting).

rosette and are 1 to 2 m in length and up to 15 cm wide. They bend inwards slightly and are dark green in colour. In its native country the sisalagave flowers around its 15th year. It develops an inflorescence up to 10 m in height which bears numerous flowers arranged in clusters. The plant attains an overall height of approximately 2 m.

Habitat: The sisalagave needs a place in direct sunlight if it is to thrive. A position near a house wall reflecting heat would be beneficial. It must also be protected from rain. In winter it is content with a fairly light situation and the temperature can be allowed to fall below 10°C on the odd occasion. The maximum temperature during the cold season should not exceed 15°C.

They must, however, be carried as hand luggage as they could suffer from the cold in the aircraft hold. Once planted, care is as for older specimens.

Pineapple
Ananas sativus
Family: Bromeliaceae

Pineapples are usually obtainable over here from May until August.
Origin: Tropical South America. Many cultivars are grown throughout the tropical latitudes.
Habit: The pineapple forms a funnel-like rosette of leaves. These leaves are linear, with entire or serrate edges according to variety, rigid, up to 90 cm long and 4 to 5 cm wide. They are covered by greyish water-absorbing scales on the upper surface. By means of these the plant is able to absorb moisture from the air. From the central funnel the main stem emerges which is c. 30 cm long.

Since the pineapple is self-pollinating it does not require the presence of other plants of its species. A mature plant will have a diameter of 1 m or more so enough space must be provided. During fruit formation a fleshy syncarp develops above the fruit; this is a dense, crown-shaped tuft of leaves and is used for the first propagation. For future propagations we can use the shoots from below the fruit, which are stronger. After harvesting the mother plant dies off.
Habitat: Full sunlight at temperatures between 25 and 30°C. In Europe we allow a resting period during winter. Reduce the watering drastically and the soil should remain only slightly moist. It may be allowed to dry out for brief spells. At the same time, as much light as possible must be provided.
Soil: Commercial potting material mixed with sharp sand. It is important to avoid wetness, as pineapple will develop root rot very quickly. Therefore do not use much peat in the soil as this is water retentive.
Watering: Regular watering with lime-free water is needed from the stage at which roots start to form. It will react quickly to too high a water content in the soil. Dry periods will disturb it less, for in common with many other species of the Bromeliaceae it is able to use the moisture in the air by means of the afore-mentioned scales. We must see that the plant is kept in an atmosphere of 60% humidity.
Feeding: During the growth period, which in Europe is the summer, a manure feed is necessary roughly twice a month. The plant has no particular decorative value and is grown principally for its fruit. It therefore needs a feed with a high phosphate content, which will encourage fruit formation. I use Guana, available in the trade. Simply sprinkle a pinch on the soil and water immediately.

To grow a
pineapple plant
and bring it to
bloom needs
really green
fingers.

Normally the plant is 3 to 4 years old before it is ready to flower.

Maturing, harvesting: After 3 years at the earliest the flowers develop on a stem c. 30 cm long. Without any need for further attention on our part the fruit will follow. This will ripen in 3 to 4 months. The pineapple can flower during any month of the year if it has a warm position in winter, even if the main flowering season remains the summer months. Similarly the fruit will ripen during any period of the year, although those which ripen in summer are much more aromatic than those harvested in winter.

Ripening is recognized by the yellowing of the fruit. At the same time they will emit an intensive scent. One must resist cutting the fruit at this time, and wait until they may be depressed by the thumb, whereby some juice might escape, for this is the moment when they are truly ripe. Depending on the cultivar the fruit will weigh up to 2 kg. They have a very aromatic flavour.

Shop fruits, harvested green because of the difficulties of shipping them, are greatly inferior in taste to the ones we grow ourselves. After harvesting the fruit, leave the mother plant *in situ*. It will gradually die off, at the same time producing lateral shoots. We remove these when they are about 30 cm long. They are grown in the same manner as the apical tuft of leaves. If they are grown under glass, shade needs to be provided only in the hottest season.

Propagation, cultivation: To start off our first pineapple plant we must obtain a fruit which we are sure has not been exposed to frost. We buy it during the summer months. Cut off the leaf crown with a sharp knife so that a wedge of fruit remains below it. Next, carefully scoop out all the fruit flesh.

The central axis, or stem, of the leafy crown should now be visible. Trim the crown to the level of the lowest small leaves, then pull 2 to 3 rows of these small leaves away in a downwards direction. The stem which is revealed shows rows of raised brown spots; the growing points of the roots. By removing the lower leaves, moisture can reach these points and root development will be encouraged.

21

We leave the leaf crown thus prepared exposed to the air for 2 to 3 days so that the cut surface can dry out and will not be a source of infection after potting has taken place. Meanwhile we must select a container. Initially a 10 cm pot is adequate. It is important for it to have a drainage hole in the bottom so that there is no accumulation of water.

When the leaf crown is ready for planting, dampen the soil slightly and press it gently in. It must then stand in a warm, light place but not in direct sunlight. A polythene bag put over the plant, tied on with a piece of string round the pot rim, will ensure the necessary moist atmosphere. Thus too much moisture will not be lost through the leaves, an important factor while there are no roots.

When the leaf crown begins to grow it is a clear sign that the roots have formed. One must not at this point lose patience, for after about 4 weeks the root system should have developed. When the young plant is obviously growing we can remove the polythene.

Cherimoya
Annona cherimola
Family: Annonaceae

The fruits are sold from January to March in well-stocked food stores.
Origin: They came from tropical America but are now grown in many warm countries.

Habit: The Annonaceae family consists of small trees or shrubs with soft, elliptic to lanceolate, finely hairy leaves equivalent in size to those of our native apple tree. The tree is evergreen, the young shoots brown. In a tree box or under glass it attains c. 2 m in height. The flowers appear on the short shoots of the branches, are largish, and white to yellowish in colour. After pollination a large fleshy syncarp develops as in the pineapple. This fruit has the size and appearance of an apple but with the difference that it is covered with scales. When the fruit is fully ripe it remains dark green.

Habitat: Since the cherimoya originally comes from the highlands of Peru, and is therefore used to intense sunlight, it needs the lightest position we can offer it. From May to September it can stand in the garden in a light situation in full sunlight. In winter it needs a light place which does not have to be warm since it is used to low temperatures in its native land. In general it is more adversely affected by temperatures over 30°C than by lower ones, which may be allowed to fall to c. 12°C for a time.

Soil: *Annona* is not difficult when it comes to choice of soil. A poorer medium, such as that used for cacti, is most suitable for it, since it will be well aerated. In soil that is too rich the plant will develop root rot,

particularly if no care is taken with water quantities.

Watering: Rain-water is best, but should be applied in moderation. The tree is fairly resistant to dry conditions, that is, it will not suffer from temporary drought, whereas excessive moisture, especially in conjunction with rich soil, will bring on root rot.

Feeding: The cherimoya does not need much feeding. It is sufficient to give it an organic feed every four weeks during the growth period.

Maturing, harvesting: After the fruits have started to form in the spring, they continue to grow during the year until the warmest temperatures have come and gone, reaching their maximum size in late autumn to winter. They remain green in colour, even when ripe. They are ready for consumption when they feel soft when pressed lightly with the finger. Nature has provided us with a real fruit cocktail, for the fruits of *Annona cherimola* taste just like strawberries and cream. The lighter the tree's position, the more highly developed is the flavour.

Propagation, cultivation: In the creamy white fruit flesh are found rather large black seeds, which can be used to obtain new stock. They are pressed into a sandy soil mixture to about 2 cm deep and kept slightly moist. They can be made to germinate in sharp sand alone, and develop, on account of the lack of nutrition, a strong root system. At temperatures of between 20 to 25°C germination will take place in 2 to 3 weeks. After single planting they need a light position where they can gradually get used to the sun.

Peanut
Arachis hypogaea
Family: Leguminosae

Roasted peanuts are well known and obtainable the whole year round but will not, of course, germinate.

Origin: Originally from South America, peanuts are grown worldwide in any regions with suitable climates.

Habit: The peanut is an annual with a maximum height of 0.5 m. The ovate or elliptic-ovate leaves are paripinnately arranged. Golden-yellow butterfly-like flowers emerge from the axils of the lower leaf-stalks. They are open for a mere few hours and wither after self-pollination. The base of the ovary then begins to grow and pushes the fruit with a so-called carpophore into the soil. After penetrating to a depth of a few centimetres the tip bends in a right angle, and here a few months later the peanut develops in its brown, straw coloured papery shell. Each shell usually contains 2 to 3 nuts. Carpophores which do not succeed in penetrating the soil do not form any nuts.

Habitat: The peanut needs plenty of warmth and sun. Temperatures between 25 and 30°C are best suited to it. Cultivation under glass all year round is recommended. It also thrives, however, in a light south or west facing window, since dry air will not damage it. It will tolerate full sunlight and in shade flower production will decrease significantly. Although in Africa it is sometimes grown as an intermediate crop with other plants, this

Left: To grow well peanuts need plenty of warmth and sunlight.

Right: The fruits of *Arbutus* have only a superficial resemblance to strawberries. They taste insipid.

is only possible as the African sun is much stronger than at temperate latitudes where the plant would not get enough sunlight.

Soil: The peanut demands a loose, rather sandy soil, so that the carpophore can penetrate it easily. The peat content of the growing medium should be very limited. Good drainage is essential and water accumulation is harmful.

Watering: It does not require much water. The soil ball should be able to dry out between the waterings. Wilted plants will revive quickly when given water. Towards the end of the growth period the soil should be only slightly moist to enable the fruits to ripen unimpeded.

Feeding: The peanut only needs a light feed every 4 to 6 weeks since as a member of the Leguminosae it takes nitrogen from the soil. All feeding may be stopped from the mid-year point.

Ripening, harvesting: When the plant yellows and shrivels it should be carefully pulled, the peanuts hanging from it, out of the ground. The nuts are plucked from their stems, but left in their shells until eaten or used for sowing.

Propagation, cultivation: In spring the peanuts are set in a poor sandy soil, with or without shell, at a depth of 1 cm. They are then lightly watered and the plant container placed in a light spot. The temperature may be allowed to reach a maximum of 30°C. During germination the night temperature should not fall below 20°C. During the first two months after germination the plants need much light and water, but when flowering begins one must reduce watering.

The peanut easily falls victim to red spider mite. As soon as there is a lightening in colour and mottled effect on the leaves, it must be treated with a pesticide, otherwise the plant will quickly succumb. Peanuts intended for sowing must be kept dry throughout the winter as they readily germinate in damp conditions, even just in high air humidity.

Strawberry tree
Arbutus unedo
Family: Ericaceae

Several nurseries have this tree in their catalogues. The reader is reminded that the two related trees namely *A. andrachne* and *A.* x *andrachnoides* are also grown in warmer parts of England.

Origin: The strawberry tree occurs in the entire Mediterranean area, in the eastern part of which a subspecies replaces the main species.

Habit: The tree, which the amateur can grow to a height of over 2 m, has stiff laurel-like leaves. Their edges are serrate. They have a waxy upper surface and are therefore very shiny in appearance. The bark is reddish in colour. In autumn the older trees bear terminal panicles of white flowers, their pitcher-shaped corolla resembling the Lily of the Valley, which by spring have become round red, strawberry-like fruits. Older trees resemble our apple trees in form.

Habitat: In view of its native area, it is not a candidate for the hot house. Once the last frosts have passed, it will enjoy a sunny sheltered spot in the garden or on the balcony. In warmer parts of England, however, *Arbutus* is quite hardy and is frequently planted in southern counties. To obtain trees of attractive shape *Arbutus* should be placed in full sunlight. In less favoured parts of England care must be taken, however, to accustom the tree to it gradually in spring or scorching will result, and this will only heal slowly. In winter a light room is required, the temperature not rising above 10°C.

Soil: The strawberry tree is extremely demanding as to growing medium. It must be acid, and soil with lime content spells death for it. (The translator notices with some bewilderment that "Hilliers Manual" states: "Unusual for its lime tolerance".) Compost mixed with some peat and sharp sand, possibly loosened by an admixture of perlite, is ideal. It must be well aerated. A high proportion of peat is to be avoided so that there is not too much water retention.

Watering: This must be done carefully. It is essential to use lime free water or rain-water. It is equally important to see that the soil ball remains slightly moist. Drying out or alternating dry and wet conditions will certainly kill the plant. A curious fact is that older plants are more sensitive in this respect than young ones.

Feeding: The strawberry tree is modest in its demands. A manure feed can be given once a month in the growth period.

Maturing, harvesting: The conditions we can offer the plant do not favour the development of fruit, but that is no great loss as they are rather tasteless. The specific epithet *unedo* indicates this: "I eat (only) one".

Propagation, cultivation: This is not feasible for the amateur, but young plants are freely available from tree nurseries.

Ramie
Boehmeria nivea
Family: Urticaceae

Not obtainable commercially.

Origin: The ramie comes from China. In the Far East there were plantations long before cotton was grown commercially. Today it is cultivated in North Africa, North America and in parts of Asia.

Habit: *Boehmeria nivea* is a coarse perennial plant up to 2 m tall. The

Boehmeria is a fibrous plant which resembles our native stinging nettle.

alternate heart-shaped leaves are, according to variety, either white, and tomentose on the underside (var. *chinensis*) or green (var. *indica*). The flowers are monoecious, arranged in panicles, and produce numerous tiny seeds. In appearance the plant resembles our native stinging nettle to which it is also related, but it is devoid of stinging hairs.

Habitat: The ramie can be grown in a pot or tub out of doors from the end of the frosts until the return of autumn. It requires a light, sheltered spot, protected from continuous rain. It is not advisable to plant it in the greenhouse as it will become too luxuriant. In winter cut off the plants just above the root system as in dahlias. The roots can then overwinter in the dark, kept only slightly moist at 10°C.

Soil: The plant requires a very nutritive soil, which should be up to 50% loam. Tap water will in time cause damage. Avoid water-logged soil or the roots will suffer.

Feeding: To obtain the strongest possible stem with well developed leaves, regular feeding, especially with a nitrogen-rich manure, is essential. This must be given from March into September at fortnightly intervals.

Maturing, harvesting: The fibres can only be obtained by an industrial process. They are about 30 cm in length and always free of wood. They are composed of pure cellulose and are 5 times stronger than cotton fibres. They are white and very shiny in appearance and may be used for the manufacture of textiles.

Propagation, cultivation: The plant is mainly propagated by division of the roots. The seed will also germinate in a mere fourteen days if sown on loose soil. The temperature does not have to be particularly high: 15°C is sufficient. It is important not to cover the small seeds with soil.

Tea
Camellia sinensis (syn. *Thea sinensis*)
Family: Theaceae

The tea plant is only obtainable in the horticultural trade from a few

Camellia sinensis can be kept without difficulty on a window ledge.

tree nurseries, if at all. If one is acquainted with someone who travels to or who comes from tea growing countries it should be possible to obtain some tea seeds.

Origin: The starting point was the Indian subcontinent, but as early as 2000 BC the tea plant is said to have been introduced to China. It first came to Europe after AD 1400 on the trade routes via the Arabs. Today it is grown on a large scale in India and Sri Lanka; however, the Soviet Union, Africa and South America now have cultivations too.

Habit: The Chinese tea plant grows to a height of 5 to 6 m; in Assam cultivated plants reach 15 m in height. Mostly, however, the plant is grown as a shrub to facilitate harvesting. Tea has alternate, dark-green, lanceolate leaves. The leaf margins are slightly serrate, somewhat leathery, up to 9 cm long and 3 cm wide. The white to pink flowers develop from the axillary buds. They have a diameter of up to 3 cm.

Habitat: It can be grown as a pot plant in a light window, put outside in the summer and may also be cultivated under glass. It does best at a temperature of between 18 and 25°C.

In the months without an 'r' in them the tea plant requires shading if under glass. This is not necessary in the open air. In winter it should have a light position, the temperature not exceeding 15°C. It may be allowed to fall to 10°C.

A high moisture content in the air throughout the year is necessary. Overwintering in a heated sitting room, especially above a radiator, is not advisable. The China tea plant can, as a pot plant, stand temperatures near 0°C.

Soil: In the countries where it is grown commercially, tea is found in the most varied soil conditions; from sandy to loamy soil, even in stony territory. As a pot plant it should be provided with a well-aerated soil rich in humus; ideal is a mixture of standard soil and ⅓ loam. The potting material must not contain lime, for the tea plant is not compatible with it.

Watering: Since the plant must not be exposed to lime, only rain-water or treated water can be used. The plant is also sensitive to variations in the humidity of the soil, and to excessive moisture. Water only when the soil surface is dry, or even better, when it can be judged from the weight of the soil ball that water is needed. Watering should be limited in winter, but the soil not allowed to dry out completely. It will benefit from spraying with lime-free water.

Feeding: The methods recommended for the coffee plant are suitable for tea, remembering that lime-free water must be used. Feeding takes place fortnightly from the beginning of the growth period to the middle of September.

Maturing, harvesting: Black and green teas derive from the same plant; only the harvesting is different. Since special machinery is needed, the amateur cannot make his own teas. In order to obtain seeds for further cultivation, one must have at least two plants of different sexes, since the tea plant is dependent on cross-pollination. Only half the flowers will produce fruit, however. It is twelve months after successful pollination before the seeds are ripe. When the capsule is ripe, it bursts and the seeds may be extracted.

Propagation, cultivation: On receiving the tea seeds, remove them from their seed case. They resemble a hazelnut in shape and size. To test for viability, put them into warm water. Those which sink are the

ones to be sown. Ripe, healthy seeds germinate after 2 to 4 weeks. Set them in peaty soil 1 cm deep and leave them in shady conditions at 20°C. When the first regular leaves appear the seedlings may be replanted, taking great care not to damage the roots.

Queensland arrowroot
Canna edulis
Family: Cannaceae

Starch is obtained from the tubers of the plant and sold as "Queensland arrowroot" starch.

Pickled in vinegar
the unopened
flower buds of
Capparis spinosa
are the capers of
commerce.

Origin: This *Canna* species comes from Central America and the northern parts of South America and is today grown commercially in Queensland (as its name suggests), Hawaii, Central America and the Pacific Islands.

Habit: *Canna edulis* resembles *Canna indica* sold here as a decorative plant. It reaches a height of c. 2 m and has typical *Canna* flowers. These are red and yellow and much smaller than those of the garden *Canna*. The leaves grow out of the stem which is formed by densely overlapping leaf-sheaths. They look rather like those of the banana, are a fresh green in colour, 12 cm wide and up to 40 cm long. Tuber-like, reddish rhizomes with a high starch content are formed in the ground.

Habitat: *Canna edulis* can be grown in the living room or under glass. It may be moved to a sheltered outdoor spot in warm weather. Temperatures over 20°C and sunlight are essential for growth. In autumn cease watering and the plant parts above ground will wither. The rhizomes (storage organs), still in the pot, will overwinter at temperatures not lower than 10°C. Planted in the greenhouse, *Canna* will form imposing bushes which flower continuously. Kept warm and light under glass it will also flower in winter. Each new stem formed will flower after attaining maximum height.

Soil: *Canna* is happy in any humus soil enriched with peat. A 20% admixture of sand is advisable since the rhizomes will rot if constantly too wet. An even moisture is desirable. Water must not be allowed to accumulate at the bottom of the pot.

Watering: *Canna* makes full use of the water content of the soil, so the latter may be allowed to dry out occasionally between watering. The plant will do better on lime-free water and an even soil moisture. By ceasing watering from October you can cause the aerial parts to die off, should you not wish to cultivate it during winter. Kept in dry conditions at not less than 10°C the rhizomes will survive for the following year.

Feeding: cf. *Ipomoea batatas* (p. 75).

Maturing, harvesting: The rhizomes contain starch. It is obtained in the same way as in the manioc. Starch from *Canna edulis* is also extracted on a commercial scale through an industrial process.

Propagation, cultivation: *Canna* is always self-pollinating. If a great many seeds are required, assistance may be given by the paintbrush. After successful pollination has been completed, the green, warty capsule grows. When it turns brown in colour, the black, round seed is ripe. It is usually about 3 mm in size.

As long as light and warmth can be provided, it is possible to sow at any time of the year. The seeds receive only a light covering of soil,

are then watered, and put in a warm, light place. Germination ensues within 4 weeks. The plant may, of course, be propagated through division of the rhizomes. If mature plants are split in this way in spring, there should be enough material for this purpose.

Caper
Capparis spinosa
Family: Capparaceae

It is only possible to obtain pickled capers in shops. They are used as a condiment.

Origin: The caper originates from the Middle East but was introduced into the Mediterranean area by the Romans. Production today is focused on the South of France and North Africa.

Habit: The caper plant is a squat shrub hardly more than 1 m tall, with long, slender branches. These bear elliptic or elliptic-ovate succulent grey-green leaves. The stem is also succulent and gets thicker as the plant ages, so that old plants have a vague similarity to pollarded willows. The flowers are large and possess four reddish-white petals and numerous anthers with dark red filaments. The club-shaped seed capsule develops out of the long-stalked ovary.

Habitat: The caper demands a place in direct sunlight. Since it tolerates dry air, it may also be kept indoors in a south facing window. It will benefit from a spell in the open air in summer. It is, however, so adaptable that it may also be kept under glass in air with a humidity of over 80%.

High air humidity will result in

dark green leaves, since in these conditions they do not develop a waxy covering, not needing to protect themselves against water evaporation. In winter the plant does not require temperatures above 12°C. They should not, however, fall below 8°C.

Soil: Since in their usual habitat the roots often grow among stones or hang over walls, an indication of their need for air, the soil we provide must be very light. Cactus potting mixture, mixed with 50% sharp sand, would be best for this plant. Any mixture used must ensure good drainage.

Watering: In the long term the plant will only tolerate lime-free water. Waterlogging will cause immediate root rot. Watering does indeed present us with some problems. Too much water given on one occasion in combination with dull weather, and the roots start to decay. In my experience it is better to wait until the leaves show signs of wilting before giving the next watering: the plant will recover very quickly. It must be remembered that the caper is a succulent plant and is therefore adapted to limited quantities of water. In winter the soil ball should be kept only slightly moist.

Feeding: A mixture of mineral fertilizer and cow manure solution should be given every 4 weeks from March to December to encourage rapid growth. The exclusive use of mineral fertilizers is not recommended since the danger of mineralization of the soil is high when plants are grown in pots and the roots of the caper are very sensitive.

Maturing, harvesting: The caper will flower for the first time after 2 to 3 years. Cutting back in the autumn will encourage flowering. In summer, during the flowering period, pick the buds daily, as soon as they reach pea-size. Dry them from 2 to 4 hours then put into vinegar. Change the vinegar twice at 8 day intervals and then the capers are ready for use.

Propagation, cultivation: The caper is self-pollinating. As soon as the pollen is ripe, that is, visible as powder, it should be transferred to the stigmas with a soft paint brush. Within 4 months the fruit capsule will have developed. When the colour changes from green to brown, the seeds are ripe and the capsule can be removed. In the following spring they are sown in the soil where the mother plant is growing. The seeds are set about 5 mm deep and the container is then placed in full sunlight. Germination follows between 14 days and 4 weeks at temperatures of 20 to 25°C.

The caper bush can also be propagated by layering. In spring, suitable branches are bent down and covered with soil; the tip of the branch remains exposed. The leaves are removed from the part of the branch in the soil. When roots have formed after a few weeks, the

Habit: The pawpaw has a 2 to 6 m tall, fleshy-woody stem, which is crowned by a tuft of long stemmed, palmately divided leaves. When they fall, they leave behind large scars. On male trees, long-stemmed panicle-like inflorescenses with white funnel-shaped flowers grow from the leaf axils. The female trees (more correctly, shrubs) bear larger, short-stemmed, somewhat waxy flowers.

Recently cultivars with bisexual flowers (e.g. 'Solo') have been bred. They develop pear-shaped fruits varying from fist to melon size. When ripe they are mottled light or yellow green in colour. The orange coloured flesh has a pleasant melon-like taste, perhaps too much on the sweet side since fruit acids are totally absent. In the hollow centre of the fruit are the numerous black seeds which cannot be eaten. The plant grows for a few years in the areas of cultivation, then dies.

Habitat: As an inhabitant of the tropics the pawpaw requires a warm, light position under glass throughout the year. It will not survive the winter in a heated living room as the air will be too dry. Temperatures of 25°C in summer and not below 18°C in winter suit the pawpaw best. The humidity should not fall below 60%.

twig can be cut from the parent and planted separately. Stem cuttings may also be made. These are cut from new growth and should be about 10 cm long. Plant them in sharp sand. Kept warm, slightly moist and light they should have grown roots by late summer.

Pawpaw
Carica papaya
Family: Caricaceae

The fruits are easily obtainable in markets and in the fruit sections of supermarkets. The pawpaw is also known as the papaya.

Origin: It originates from Central America but was introduced elsewhere from there and is now a common feature of tropical regions throughout the world.

Soil: The pawpaw is happy in most soils, but they must be well drained and aerated. Constant wetness and waterlogging will soon kill the plant.

The kapok tree
will only develop
its fibrous fruit
hairs in its natural
habitat.

Watering: The pawpaw needs a lot of water. Nevertheless waterlogging must be avoided or root rot will set in. The aim should be an even moisture of the soil ball. Cut back greatly on watering in winter. Tap water will do damage in the long term. In winter the water should be tepid.

Feeding: Give a feed every four weeks during the growth period. Organic fertilizer is preferable to the chemical sort which will lead to mineralization of the soil.

Maturing, harvesting: In the tropics the shrub bears fruit all year round; under glass only in the warm season. After successful pollination, the following months see the development of a melon-sized pear-shaped fruit according to cultivar. It is ripe when the skin has gone yellow and may be depressed by the finger. Fruit harvested in the warm season is tastier than that picked in winter. It must be picked when fully ripe as it will not ripen once plucked from the plant.

Propagation, cultivation: Lay the washed seeds taken from the fruit on humus soil and cover well to about 1 cm. Kept light and at 20 to 25°C germination will follow in 2 to 4 weeks.

Four weeks after germination the young plants may be transplanted, taking care not to damage the root system. If you have not sown the cultivar 'Solo', you will need to identify the male and female plants: this can only be done when they flower. In good growing conditions the plants will flower after approximately 3 years. Washed and dried, seeds may be kept in jars for years.

Kapok
Ceiba pentandra
Family: Bombacaceae

The kapok is a fibre plant, its fruit hairs being used for mattresses and cushions as well as insulating material.

Origin: Native to the warm zones of South America but grown today in the tropics worldwide.

Habit: The kapok tree is one of the largest trees of the tropical forest, with powerful, widely flaring buttresses. It attains a height of up to 50 m. The branches are high, widespreading and arranged in several separated stories. In its early stage the trunk is armed with short, conical spines although there are also forms which do not have this characteristic. The stalked leaves are palmate with usually 7 lanceolate leaflets about 2 cm across. These are shed in the dry period. At the same time the showy flowers appear in clusters on the outer axils of the lateral branches. They are white to yellowish, slightly purple at the base and are wind pollinated.

Leathery brown fruit capsules develop from the ovary. They are 10 to 20 cm long and spindle-shaped. They contain the pea-sized black

seeds in the centre of silky, shiny fibres 1 to 2 cm long which are used as a mattress filling, and more importantly, since they are hygrophobic, as insulating material against damp. They are not suitable for spinning.

Habitat: The kapok needs a position under glass all year round at temperatures of 25 to 30°C in summer and not below 15°C in winter. The new growth must be protected from too much sun, but otherwise full light is tolerated. In winter it needs as much light as possible. It sheds its leaves every year in the winter, which corresponds to the dry period in its native country. If placed in darker conditions when leafless, it can fail to produce new growth in spring.

Soil: Kapok needs a well-aerated humus soil mixed with sand. It should be slightly acid. Drainage is very important since the plant is extremely sensitive to water accumulation.

Watering: The kapok needs plenty of water in summer, but constant wetness of the soil ball must be avoided. During the leafless period a slightly moist soil is all that is necessary. The water should only be increased when there is evidence of new growth.

Feeding: The plant needs a manure feed every three weeks so that it can achieve the rapid growth which is natural to it. It can attain a height of 2 m within 3 years. Inadequate feeding will result in a starved appearance which has nothing in common with the natural form of the plant. Do not, of course, feed during the leafless period.

Maturing, harvesting: The kapok

will not flower in our greenhouses.
Propagation, cultivation: Only fresh seeds will germinate. Place them in lukewarm water, 30°C, for 48 hours. Any floating on the top after this period are infertile. Set the seeds c. 1 cm deep in loose soil and keep in the shade at a temperature around 25°C. They will germinate after about 14 days. The seedlings can be thinned out as soon as the first palmate leaf appears.

Kapok can also be propagated by taking stem cuttings. With the exception of the apex, all parts of the stem will grow roots within 3 weeks if kept in warm, shady conditions.

St. John's bread, carob
Ceratonia siliqua
Family: Leguminosae

The dried long pods of this plant, sold as St. John's bread, are obtainable all year round.
Origin: The tree is a native of the Middle East and is now found particularly in the Mediterranean region.

Habit: In some parts of the Mediterranean region these towering trees with their mighty canopy stand out in an otherwise bare landscape. *Ceratonia* has evergreen, palmately divided hard leaves. The new growth is red and soft. Flowers cannot be expected in our climate. They are suspended on extraordinarily long stalks and have showy purple stamens. During the flowering period the tree emits a foul smell.

The well-known brown (originally white) hard-shelled pods contain seeds which are so consistent in weight that they are still sometimes used, especially in the Middle East as units of weight. The word "carat" is derived from the plant's name. If well cared for, the tree can reach a height of 2 m or more.
Habitat: The small, round, very tough leaves of this tree tell us that it is adapted to the heat. For this reason we shall give it as light and sunny a position as possible. After the last frosts it can be put outside and stay there until the onset of the next autumn frosts. In winter it

St. John's bread is available in most healthfood shops.

needs a cool light place. The temperature may be allowed to fall to 10°C.

Soil: The plant needs a poor soil which drains well, such as potting mixtures for cacti. It should have some lime content. In nature it favours dry, stony slopes.

Watering: As already mentioned, the carob is very resistant to dry conditions. Restraint must be the watchword when watering. No *Ceratonia* ever died from dryness but many have succumbed as a result of well-intentioned over-watering. Rain-water is good, but, if lacking, tap water may be given.

Feeding: Since the plant grows slowly, one must go carefully with feeding, otherwise it will become too luxuriant. One feed at the beginning of the growth period is enough. Since it develops an extensive root system it should always have a large container, providing access to a larger source of nutrition. The soil should be only slightly moist in winter, otherwise the carob, as a plant resistant to dryness, will react with root rot, especially if fertilizer salts remain in the soil.

Maturing, harvesting: Since the tree will not bear fruit in our climate advice would be superfluous.

Propagation, cultivation: Remove the seeds from the dry pods and lay them on a suitable soil (see above). Cover with a small amount of soil; about as deep as the seeds are thick. Moisten slightly, cover with polythene and keep light and warm (20 to 25°C) but protected from direct sunlight. When the seeds germinate after about 3 weeks, the polythene must immediately be removed. After thinning, place the young plants in the sunlight.

Cinnamon
Cinnamomum verum (C. zeylanicum)
Family: Lauraceae

Only cinnamon products in the form of powder or sticks are commercially available, not the actual plant.

Origin: Cinnamon grows wild in India and Ceylon. Today it is cultivated everywhere in Asia where the climate is adequately warm and humid.

Habit: In natural conditions it is a 10 to 15 m tree with fresh green, elliptic, opposite, relatively tough leaves. In cultivation it is kept in shrub form in order to facilitate the collection of the bark which is the actual commercial product. The flowers are small, white and arranged in umbels. Berries follow flowering.

Habitat: In view of its natural habitat, namely the warm humid regions, here cinammon is kept all year round in the greenhouse. Cultivation in the living room or outdoors, even in summer, will not be tolerated by the plant. Greenhouse

temperatures of 30°C or more are suitable and they should not fall below 18°C in winter. High humidity is essential. Daily spraying with soft, tepid water will encourage healthy development.

Only in the hottest part of the year is it necessary to provide slight shade, otherwise cinnamon enjoys maximum light. It must be borne in mind that light is more intense in its homelands than here, so that when we read that it grows in the shade there we must not assume that that is advice for our climate.

Soil: The cinnamon tree requires a humus soil that drains well. Up to ⅓ sharp sand (bird or aquarium sand) may be mixed with the soil, which may be a standard potting mixture. It is important for superfluous water to drain away quickly and for the roots to be well aerated. In peaty soils, which decompose too fast and retain water, the root system will soon rot.

Watering: The cinnamon tree requires a slightly acid soil, so rainwater should always be used. It needs plenty of water in summer, and must be kept only slightly moist in winter. Waterlogging is fatal.

Feeding: Every four weeks in the period from April to September is sufficient. A mixture of cow manure and chemical fertilizer is advisable.

Maturing, harvesting: The commercial product of the cinnamon tree is the bark, which is obtained

In contrast to the lemon, the fruits of the lime will remain green when mature.

from the twigs after 2 years' growth. The twig is stripped using longitudinal and cross cuts. The bark ferments overnight in the warm moist atmosphere. The outer layers are then scraped off. The inner layers are now dried in the sun turning their characteristic cinnamon-brown colour. Cinnamon cultivated under glass is not nearly so aromatic as that which grows in its natural habitat.

Propagation, cultivation: Seeds may be used in the usual way but must not be more than 4 weeks old. Generally cinnamon is propagated by means of cuttings. The cuttings, c. 15 cm long, taken from the apical part of the twigs are simply set in sand, given a light watering and a plastic covering, then kept in the shade at temperatures between 25 and 30°C. Cuttings taken in May will have rooted by July.

Lime
Citrus aurantiifolia
Family: Rutaceae

The cultivar usually sold in fruit shops is 'Tahiti', but it is seedless. Since plants are not obtainable in the trade we must try to obtain one from the native country.

Origin: Lime varieties are mainly grown in the Caribbean and in Mexico.

Habit: *Citrus aurantiifolia* forms small shrubs with thorny branches. The leaves are smaller than those of other *Citrus* species and their stalks are not winged. The flowers are also smaller and white with yellow stamens; they can be self–pollinating.

The fruits of the cultivar 'Tahiti' are ovoid and about hen egg sized; those of the Mexican variety are globular. The former cultivar is always seedless; the second produces seeds, but is hardly ever sold in the fruit trade. The fruits are grass-green, even when ripe. The well-known "lime juice" is a product of this fruit.

Habitat: The lime requires a warm, humid position the whole year round, growing, as it does, in a tropical climate. In summer it can hardly be kept too warm and in winter the temperature should not fall below 18°C. It can also tolerate full sunlight under glass in summer. In winter, place it, in its pot, on a heating pipe in the greenhouse, so that temperatures of 18 to 20°C are maintained in the soil ball. It will not grow in the living room, nor should it be kept outside. In view of its origin, high air humidity is necessary.

The Seville orange adorned many an orangery in the past. The fruits are unsuitable for eating fresh but are the basic ingredient of marmalade.

Soil: The lime is not very demanding as far as soil is concerned. A well-drained, humus soil with an admixture of loam is ideal.

Watering: Care must be taken that there is an even moisture content in the soil. Overwatering, in conjunction with low temperature, can spell death for the plant. The more loam there is in the soil, the more carefully you must water. All citrus plants have leathery leaves, an indication that they can manage on a moderate amount of water, although if the soil dries out damage will be caused. In winter the water, preferably rain-water, should be slightly warm.

Feeding: A feed every 4 weeks from the onset of growth till the end of September is beneficial. The prolonged feeding period derives from the fact that, if well cared for, the plant will produce many fruits and consequently needs plenty of nutrients so that it does not shed them. This applies to all *Citrus* species.

Maturing, harvesting: The first limes can be picked from Christmas onwards. The fruits may be left on the tree until thay fall off by themselves. Even when ripe they are very sour and are generally only grown for juice production.

Propagation, cultivation: You could try to propagate the lime, like other *Citrus* species, by making cuttings, but the results are not very impressive. Should you obtain a cutting from a botanical garden, it must be transported in humid and warm conditions. Set in sterilized sharp sand and kept shady at 25 to 30°C, not forgetting the obligatory polythene bag, root formation might follow. Better results will be obtained by grafting onto a suitable species, e.g. *Poncirus trifoliata*.

Seville orange
Citrus aurantium subsp.
aurantium
Family: Rutaceae

Seville oranges are readily available from fruit shops. They are used in marmalade and in liqueur manufacture.

Origin: A native of northern India, the Seville orange was introduced into Europe very early and was often grown in the orangeries of the aristocracy.

Habit: The Seville orange can take the form of a large shrub or a small tree a few metres high. It also has the relatively leathery leaves characteristic of all *Citrus* species. They contain essential oils. The flowers are small, white and scented. After self-pollination small, orange-coloured fruits with a rough, pitted, very thick peel develop during the course of the year. The branches and twigs have, depending on the individual specimen, a few or many spines.

Habitat: Since the Seville orange is not too sensitive to the cold, it will do well in the open air in a sunny warm position after the May frosts until the autumn frosts. It should be protected from wind. It is toler-ant of temperature changes and it can hardly be too hot for it in our climate. A light position is necessary in winter, with temperatures not exceeding 5°C if possible.

Soil: It needs the same soil as other *Citrus* species, even though it will thrive in rich soil.

Watering: The plant demands an even moisture content in the soil. It will not suffer so readily from dryness as other *Citrus* species. In winter the soil must be kept only slightly moist. When the leaves begin to curl slightly, water should be given.

Feeding: Feed with restraint, otherwise the plant will become too luxuriant and demand more space than the amateur can provide. Since it never produces many fruits when grown in this country, it is not necessary to feed it copiously.

Maturing, harvesting: Yields of several kg of fruit can be expected annually. It is not a suitable dessert fruit but excellent for marmalade. The thick rind is also used for the preparation of dried or crystallized peel.

Propagation, cultivation: Plants and seeds are difficult to obtain, but when a grafted plant dies, you should, if possible, allow the stock to shoot. Many grafted varieties are based on *Citrus aurantium*.

Myrtle-leaved mandarin
Citrus aurantium var. *myrtifolia*
Family: Rutaceae

As a fruit it is not obtainable in the trade. Plants are occasionally offered in horticultural catalogues.
Origin: The myrtle-leaved mandarin can be cultivated anywhere where other mandarins are to be found. It is not grown commercially.
Habit: The plant forms a small shrub with densely arranged leaves. They are elliptic, only 2 cm long and equally broad. The small flowers are white, scentless and self-pollinating. The yellow-red fruits are tennis ball size and hang thickly together from the branches.
Habitat: cf. Lemon (*Citrus limon*) p. 43.
Soil: cf. Lemon p. 43.
Watering: cf. Lemon p. 43.
Feeding: cf. Lemon p. 43.
Maturing, harvesting: The fruit is ready around Christmas and is eaten raw.

Cultivars which in some cases are of considerable commercial importance are hybrids of the mandarin and grapefruit, known as 'Orlando' or 'Mineola'; and of the tangerine and orange ('Clementine'). The best known cultivar is 'Temple'.

It is important for the citrus plant enthusiast to know that clementines are self sterile, i.e. they need the pollen of other citrus species, with the exception of the orange cultivar 'Washington navel' which has sterile pollen. This pollen cannot, therefore, fertilize its own cultivar or the flowers of other *Citrus* species.
Propagation, cultivation: By grafting on to suitable species.

Lemon
Citrus limon
Family: Rutaceae

Fruits are obtainable all year round, and plants occasionally.
Origin: A native of China, the lemon was brought to Europe by the Arabs. Today it is cultivated in all subtropical regions of the world.
Habit: In size it' is between the orange and the mandarin tree. It, too, has leathery leaves supplied

with oil glands. The flowers are pinkish and have a strong fragrance. As with all *Citrus* species, its flowering period is spring in our climate. It continues to bear flowers after the main flowering period is over, so that we can observe ripe and unripe fruit and flowers on a tree at the same time.

The tree renews its leaves throughout the year. Some branches become quite bare, but since these are the branches most likely to flower, do not cut them off. Too radical pruning will not result in increased flowering, but the tree will produce many epicormic shoots.

The fruits may remain on the tree up to half a year without loss of quality.

Habitat: The lemon tree prefers a light, sunny position. In our climate it is best kept throughout the year under glass, as this will encourage a good flavour. If you are less concerned about obtaining large, wholesome and tasty fruit, then your lemon tree may be kept in the open air in a sunny but sheltered spot, since it is sensitive to strong wind. In winter keep it in a light cool place at not more than 10°C.

Soil: cf. Grapefruit (*Citrus* x *paradisii*) p. 47.

Watering: To obtain a good harvest copious water must be provided during the growth period. Here again, rain-water is best. Between waterings, however, the soil ball should be allowed to dry out somewhat, so that it remains only slightly moist. Care must be taken, for dryness will bring a swift reaction from the tree and it will shed its fruit. In winter the soil must be kept slightly moist, just sufficiently to prevent leaf curl and subsequent leaf loss.

Feeding: It is obvious that a plant which is required to give a fruit crop will need plenty of nutrition. For this reason give it a feed in the form of liquid manure up to twice a week in the growth period. The lemon tree is also sensitive to an accumulation of mineral salts in the soil, so organic fertilizer is recommended. Feeding must cease by the end of August at the latest even if the tree is to go on growing, otherwise next season's flowers are at risk.

Maturing, harvesting: The first ripe fruits are generally to be harvested around Christmas. Since they may, however, be left on the tree for up to 6 months, one has freshly picked fruit at one's disposal up to mid summer. A 6 to 8 year old tree can bear from 15 to 25 kg of fruit. These will be far superior in aroma and juice content to those in the shops. The largest lemon I have ever harvested weighed 350 g.

Propagation, cultivation: The lemon plant is usually grafted. Cuttings c. 20 cm long which are not fully mature may be taken from recent growth. The cut should be made just below a bud. The ends

The calamondin is frequently sold as an ornamental plant.

are dipped in rooting powder, set in a humus soil, put in a plastic bag and kept shady and warm. Not every cutting will take, but after c. 8 weeks some should be showing the first roots. Cuttings taken in late summer may not show growth until the following spring. As long as the cutting is green it is alive and can start to grow. Lemon trees are frequently on sale in garden centres.

Calamondin
Citrus madurensis
Family: Rutaceae

This is obtainable as a decorative plant in garden centres and flower shops throughout the year.
Origin: The Far East.

Habit: This is an orange-like plant of maximum height 1.50 m with small, orange-coloured, plum-sized fruits. The leaves are shiny, green and ovate. The flowers are white, small and fragrant. The plant flowers and fruits prolifically. It will bear ripe and unripe fruit and flowers simultaneously.
Habitat: The calamondin requires a light, but not too warm position. It can be put outside after the May frosts and left there until the return of the first night frosts. The sun will not harm it in the open air. In winter it needs a light but cool spot. The heated living room would be too warm. A cool staircase or unheated room would be suitable until it is once more placed on the balcony or in the garden in spring.
Soil: The calamondin needs a loose,

aerated humus soil. Loam content should be avoided. Leaf litter, but with little peat (which is water retentive), is the correct potting material.

Watering: I have observed that *Citrus madurensis* is more sensitive than any other *Citrus* species to soil that is too wet, therefore caution is advised when watering. Soil that is slightly moist all year round is best suited to the plant. You are advised to use rain-water since the plant's leaves readily turn yellow. If in doubt as to whether to water or not, don't! Drought damage is easily repaired, but too much water will lead to root rot and the death of the plant.

Feeding: To get a good fruit crop, a weekly feed from spring to early September is advisable. To give a rough idea, I put a fist sized quantity of cow manure into 5 l of water and leave to stand for 2 days. Stir well before using. This mixture has not yet harmed any plant, and avoids the root damage which results from persistent use of chemical fertilizers. This recipe for feeding keeps any offensive smell to a minimum.

Maturing, harvesting: The small fruits may be harvested all year round. They taste very bitter and are not palatable eaten raw. A good jam may be made with them.

Propagation, cultivation: It is propagated by grafting. Since, however, young plants are readily available in the trade, there is no problem with replacement. In my experience it is not easy to grow the calamondin and a great deal of intuition and observation is required.

Shadok or pummelo
Citrus maxima (C. grandis)
Family: Rutaceae

Origin: It is found in many forms, particularly in South East Asia. Its fruits are sold in good fruit shops but plants are seldom available in the trade. Occasionally young plants may be offered in horticultural catalogues along with other *Citrus* species.

Habit: In its natural habitat it becomes a 6 to 8 m tree. Its leaves are the largest of all *Citrus* species; 12 cm long and up to 8 cm wide. They have a very wide leaf-stalk, which can lead one to believe it is a second leaf. The flowers, appearing in spring as all *Citrus* flowers do, are about the size of a 5p coin and feel like hard wax. They are self-pollinating. The fruits, the size of a baby's head, develop from the ovary and are yellow when ripe. The peel is very thick and easily detachable from the flesh, which has a slightly bitter taste, somewhere between that of grapefruit and blackberries.

Habitat: In our climate the plant must be kept in the heated greenhouse. It will not benefit from a spell in the open air in summer.

Growth is too often inhibited. Since it requires high humidity it cannot be kept in the living room. Temperatures should rise to 30°C and more in summer, but the plant should be slightly shaded in the hottest period to avoid scorching of the leaves and fruits. In winter the temperature should be not allowed to fall below 15°C, and the soil ball kept only slightly moist. It needs plenty of light all year round.

Soil: The plant will grow happily in a potting material as described for *Citrus aurantiifolia*. To obtain maximum sized fruit it should be planted in the centre of the greenhouse.

Watering: In its natural habitat each rainstorm supplies ample rain-water, and it will survive on this for 2 to 3 weeks. As a pot plant we must ensure it constant slight soil moisture. Daily spraying will be of great benefit to the plant.

Feeding: cf, *Citrus aurantiifolia* p. 40.

Maturing, harvesting: The fruits are ripe when they become canary yellow and the peel yields when lightly pressed. They may be eaten raw. They will only acquire the desired sweetness when they have had enough sun.

Propagation, cultivation: It may be grafted on to the bitter orange (*Poncirus trifoliata*), using a T-cut, in August. Otherwise keep your eyes open for offers in horticultural catalogues.

Citron
Citrus medica
Family: Rutaceae

The fruit is rarely seen on sale.

Origin: Its original home is in Asia. Today it is a common sight in Greece and Sicily.

Habit: The citron tree resembles the lemon tree in size and shape. The fruits, however, have a thicker, wartier peel. The amount of fruit flesh is minimal.

Habitat: cf. Lemon p. 42/43.
Soil: cf. Lemon p. 42/43.
Watering: cf. Lemon p. 42/43.
Feeding: cf. Lemon p. 42/43.
Maturing, harvesting: The ripe fruits are harvested in the winter months. They are not suitable for

eating, but candied peel is made from them by first dipping the peel in salt water then immersing in sugar syrup.

Propagation, cultivation: It is not possible to start a collection other than by obtaining young plants from the growing regions.

Grapefruit
Citrus x paradisii
Family: Rutaceae

The fruits are obtainable all year round.

Origin: The grapefruit was first grown in the West Indies and is now cultivated worldwide in subtropical regions.

Habit: This *Citrus* species can attain a height of 10 m, but in our climate will seldom exceed 3 m. It has shiny leaves provided with oil glands, and a widened leaf stalk looking like a second leaf. In a suitable situation this species will flower all year round. The flowers, white or with a hint of pink, are strongly scented, and hang in grapelike clusters (hence the name) at the end of the youngest shoots. They produce so much pollen that one flower is often sufficient to pollinate most of the flowers on the tree. You may remove blossom without fear, since the tree will only retain two or three pollinated flowers per cluster, which in view of the fruit size is understandable.

Habitat: The grapefruit tree can only be grown under glass in our climate. It may be placed outside in the warmest months, but in view of its high temperature requirements, this could affect the development of the fruit. Its position should therefore be light and sunny. In winter it requires a light place and the temperature should not fall below 10°C.

Soil: The soil should, as for most *Citrus* species, be a humus soil and contain a little loam. In view of the size of the plant, a tub rather than a pot is recommended.

Watering: Bearing such large fruits, the grapefruit tree needs plenty of water during their development, as the fruit-flesh will become dry if it goes short. Rain-water is best. Between waterings however, let it become drier, to avoid damage to the roots.

Feeding: To produce good, heavy fruits, the grapefruit needs plenty of nutrients. The plant will respond with stronger growth to a solution of organic fertilizer every eight days. Chemical feeding is not advisable as the plant is sensitive to mineral salt accumulation in the soil. Cease feeding at the end of August to stop further wood growth. If the plant continues to grow during winter, and this applies to all *Citrus* species, next year's flowering is adversely affected.

Maturing, harvesting: cf. following species.

Propagation, cultivation: The information for the mandarin (below) also applies to this plant.

Mandarin
Citrus reticulata
Family: Rutaceae

Obtainable in fruit shops in autumn and winter.
Origin: Natives of South East Asia, all varieties of mandarin are today grown world-wide.
Habit: Compared with the orange tree, the mandarin is smaller and has smaller willow-like leaves. The branches are somewhat pendulous. The flowers are correspondingly smaller, white and also fragrant. The mandarin is, like the orange, self-pollinating. Since it rarely attains more than 2 m in height in our climate, it resembles a shrub rather than a tree.
Habitat: Light and sunny, but only outdoors in the warmest summer. Since it is much smaller than the orange tree, it may easily be moved about in its tub. In winter it requires a light, but relatively cool place (not over 10°C). The leaves are shed and renewed throughout the year if the plant is not suffering from disease or root rot. Do not remove bare branches, as these are likely to flower, as long as they look green and healthy.
Soil: The mandarin prefers a lighter soil than the orange; it will also do without the loam content. Well aerated soil is important. Caution is required with regard to peat content, since the mandarin does not like prolonged periods of wetness and will develop root rot.
Watering: Rain-water is again the best. Between waterings, not only the soil surface but the entire soil ball should be allowed to dry out somewhat. This can be tested by feeling the weight of the plant plus tub.
Feeding: Regular feeding is necessary to obtain a good fruit crop. An organic feed may be given every 8 days in the growth period. If short of food, the mandarin will drop its fruit. Keep the soil ball slightly moist in winter.
Maturing, harvesting: If successful pollination has taken place in spring, the resulting fruit may be harvested, depending on the hours of sunshine and the degree of warmth the plant has had, from November onwards. A 5 to 8 year old plant will yield c. 5 kg of fruit. The flavour and sugar content will be more highly developed if the plant has been grown under glass. Mandarins cannot be left on the tree, unlike lemons, and must be picked immediately when ripe.
Propagation, cultivation: Cuttings have not proved successful in the case of the mandarin, so grafting is the alternative. Sometimes airlayering (see p. 119) will lead to success. Study horticultural catalogues, especially the spring issues,

as a few firms offer citrus plants around this period.

Orange
Citrus sinensis
Family: Rutaceae

Oranges are obtainable in fruit shops all year round.

Origin: They are probably natives of China but are cultivated throughout the world in the sub-tropical regions.

Habit: The orange plant is a small tree or shrub from 3 to 5 m tall. The leaves are relatively hard, leathery and with numerous oil glands. The white, waxy, five-petalled flowers have a very pleasant scent. It may have spines to a lesser or greater degree.

The orange plant is almost always grafted. Depending on origin, various stocks are used, although some varieties are known to breed true from seed. After pollination, which must be done by hand if the plants are kept indoors, the fruits develop in the course of the summer. They turn orange and ripen in the autumn. If they have had enough sun they will not be inferior to those imported commercially.

The orange tree, and this applies to all *Citrus* species, should not be cut back, or at least, very cautiously, otherwise it will react by producing epicormic shoots.

Habitat: After the May frosts orange trees may be placed in a warm, sunny place and remain there until late autumn. The more sun the tree gets, the better. It must, however, receive a gradual introduction to the sun in spring. In winter keep it light and cool; temperatures may be allowed to fall to 5°C. It must on no account be kept too warm, as the short days and lower temperatures are necessary to encourage flowering.

Soil: It should have a humus soil, but mixed with one third loam. Orange plants may be grown in heavy soil but the fruits will then have a thicker peel. The heavier the soil, the thicker the peel. An admixture of lime-free sand is to be recommended, as it will provide a better aerated growing medium. The roots of the orange tree, like those of all other *Citrus* species, live in symbiosis with a fungus which enables them to make use of the nutrients in the soil. This symbiosis is only possible, however, in a well aerated soil.

Watering: The orange tree needs plenty of water during the growing period. Rain-water is best. After watering, the soil should be allowed to dry out to enable the symbiotic process mentioned above to take place. To check whether the plant is in need of water, do a weight test by lifting it in its tub. This check is necessary since the orange tree reacts badly to too much water or too lengthy

The coconut plant
is often sold in
garden centres
and nurseries in
the form
illustrated here.

periods of wetness. In winter, give only enough water to maintain a slight moisture in the soil. If in doubt as to whether or not to water, don't!

Feeding: To produce a good fruit crop, the orange tree needs a good deal of nutrients. The best results are obtained with organic fertilizers, since all *Citrus* species are sensitive to mineral accumulation in the soil.

Take caution with manure from intensive farming units, as antibiotics present can destroy our plant, just as they destroy all organisms in its soil.

Feed every fortnight during the growing period with manure dissolved in water and mixed with rainwater. Cease feeding from mid-August so that growth may come to a halt.

Maturing, harvesting: From the age of 2 to 3 years grafted orange trees will bear a greater crop each year, so that a 10 year old tree can be expected to yield 10 to 15 kg of fruit. Ripe fruits fall in autumn or winter. They can be picked when the peel has turned orange and gives to the touch.

Propagation, cultivation: Cuttings may be taken. In spring mature cuttings (recognizable by their round shape) are set in sharp sand and placed in a shaded spot at 20 to 25°C. Not every cutting will take, but after 4 to 8 weeks root development should begin. The process may be speeded up by the use of rooting powder. Air-layering (see p. 119) may sometimes succeed. Orange plants grown from pips will generally flower for the first time after 6 years, but it must be noted that you will have no information as to the identity of the other parent, or the nature of the fruit that will develop.

Coconut
Cocos nucifera
Family: Palmae

Coconuts are obtainable all year round. Germinating nuts with 1 to 2 cotyledons are occasionally sold as houseplants in garden centres. These plants may already have a height of more than 1.5 m.

Origin: Experts are not agreed upon this matter. Some maintain that it comes from South America, others are of the opinion that the South Sea Islands are its home. Today the palm is found in the subtropical belt throughout the world. It has colonized mainly coastal regions.

Habit: In its natural habitat the coconut palm forms a tall unbranched trunk up to 30 m high. At the top is a dense crown of 30 to 50 pinnate leaves which can be more than 5 m long. The paniculate inflorescences, with several thousand male and far fewer female flowers, appear in the leaf axils.

Since the male flowers open first and have withered by the time the

females are ready, pollination from another plant is necessary to obtain fruit. The coconut takes 8 to 9 months to ripen.

Habitat: This plant needs a very light position under glass all year round. Any shade during the day is detrimental. Temperatures between 25 and 30° during summer suit it best. Even in winter the temperature should not fall below 20°C, particularly in the root area.

The germinating coconut plant is often sold to us as an ideal indoor plant. This it most certainly is not.

As a result of its high light requirements it will be too dark for it in the average living room. The necessary high humidity will also be lacking. It could struggle through the summer but it would most certainly die in winter. Even in the best greenhouse conditions, the gardener is pleased when at most he keeps the damage to a minimum. Artificial light provided during the darkest winter months will make cultivation considerably easier.

Soil: The coconut palm will do best in a potting mixture which will be half well rotted leaf litter and half loam. When transplanting, one must exercise the greatest care not to detach the nut from the roots since at this stage the nutrition passes to the plant via the nut. When the pot has been topped up, gently press the soil down with a stick, since the plant appreciates firm anchorage.

Watering: The plant requires an even, slight moistness in the root area. This should be maintained throughout the winter. It is not particularly sensitive to mineral salt accumulations in the soil. If the water is not given at room temperature the plant's growth will be arrested. This temperature requirement is particularly important in winter since the fleshy roots are very subject to root rot.

Feeding: The coconut palm must not be fed too generously or it will soon exceed the dimensions of the small greenhouse. Once in spring, then again in summer, is all that is required for limited growth. We can only cope with this plant in its young form so we must keep it at this stage as long as possible.

Maturing, harvesting: Since the coconut palm does not fruit in our climate, advice would be superfluous.

Propagation, cultivation: We have only two possibilities. We either obtain a young plant from some source, or we attempt to make a coconut germinate.

In the latter case, buy a medium sized coconut in summer. You should be able to hear the milk when you shake it. Put into a damp polythene bag and hang from the roof of the greenhouse in a shady but light position. Take care that no fungal growth occurs.

Germination can take six months or more or may not occur at all. First the roots grow beneath the fibre layer and then break through it to the outside. There is no rush to plant it. Only when the nut is covered in white roots do we undertake this.

Coffee
Coffea arabica
Family: Rubiaceae

Coffee plants are available from larger horticultural establishments. Growing from seed is not feasible since the coffee bean loses its fertility after about 4 weeks and those we can obtain are generally older than this. Botanical gardens will sometimes give you a few seeds.

Origin: It is a native of Ethiopia but is today cultivated throughout the world. South America and Africa are the areas of greatest commercial importance.

Habit: It is a small tree or shrub, in nature up to 6 m high, in cultivation kept smaller to facilitate harvesting. It has opposite leaves. They are short-stalked, shiny and leathery, oblong-ovate in shape.

The best place for
the coffee plant is
the greenhouse.

The lateral branches are somewhat arching.

The pure white flowers are arranged in groups of 5 to 15 in the leaf axils of the lateral branches. They exude a jasmine-like scent. The flowering time is usually restricted to a few hours, during which time the shrub looks as if it is covered in snow. From the ovary develops a two-seeded berry, green at first and dark red when ripe. The ripening takes about 12 months in our climate.

Habitat: The coffee plant is an ideal indoor plant for the east or west facing window. In good summers it can also stand outside in a warm sheltered spot. It will grow best, however, if planted in the greenhouse at temperatures of 18°C to 25°C. Whether in the window or the greenhouse, the plant, growing in nature in forests shaded by trees, will require slight shading in the summer months. It is happy with temperatures below 20°C in winter, indeed, they may go down to 15°C. A fairly high air humidity will be beneficial. If kept indoors, the plant should be put into a cooler room in winter. It should certainly not stand on the radiator shelf.

Soil: The coffee plant needs a humus well-drained soil. Standard potting mixture mixed with one quarter peat is best. It must be on the acidic side.

Watering: To grow without setbacks, it needs an evenly moist soil ball. It must be kept drier in the lower winter temperatures. Rainwater or treated water is the only type that can be given as the coffee plant will not tolerate a lime content. Frequent spraying with lime-free water will help it to flourish.

Feeding: Organic fertilizer has proved best in my experience. Put 1 to 2 handfuls of manure into 10 l of rainwater. Stir and leave for 24 to 48 hours. Stir again and it is ready for use. Feed with this mixture every fortnight from the onset of the growth period until the end of

August. Healthy growth, dark green leaves and, when the plant is mature enough, copious flowering will result.

Maturing, harvesting: Flowering will begin when the plant is 3 years old. To obtain a good crop of fruit, it is advisable to assist pollination with a soft haired paint brush. The berry, each containing two seeds (the coffee beans), will be ripe after one year. They are then dark red in colour. If you wish to go to the trouble of harvesting and then processing the beans, remove them from the fruits, wash them, dry them, and then roast over gentle heat in a frying pan with a little oil until they are dark brown. Drinking coffee from one's own harvest is not, after all, an everyday experience. If used for germination the seed must first be extracted from the fruit.

Propagation, cultivation: The plant is grown from seeds and cuttings. Fresh seeds are taken out of the berries and the silvery skins on the seeds removed; these are a protection against water.

The same soil as that in which the mother plant is growing can be used for the seeds. They are planted flat about 1 cm deep and covered with soil. Fresh seed will germinate after 5 to 6 weeks. As soon as the second pair of leaves have formed after the cotyledons, the first thinning may take place.

Cuttings are taken from the soft new growth of the apical shoots.

They should be about 15 cm long. Place them in peaty soil and keep at 20 to 25°C in the shade and inside a plastic bag. The base of the cutting should be already slightly woody. Only the latest leaves are left on the cutting. After a few weeks it should have taken root. Only cuttings from the central shoots can be used, since only these will have upright growth. Cuttings from lateral branches will always grow laterally. These can, however, make attractive hanging plants.

Dasheen
Colocasia esculenta
Family: Araceae

Neither the plant nor its products are available in the trade.

Origin: A native of India and the islands of the Sunda Archipelago, it is now grown in the entire tropical area.

Habit: The dasheen develops a knotty subterranean rhizome. From this grow several huge long-stalked leaves, which can, according to species, be from 1.5 to 2 m long. They are dark green with obvious veins. The flower resembles that of the Arum lily. During the growing period, 1 to 3 heavy main corms are developed and several small daughter corms. These corms, rich in starch, are the useful parts of the plant. Young leaves and leaf stems are also eaten

The dasheen, a tropical marsh plant, demands a warm humid climate and must therefore be cultivated under glass throughout the year.

as a vegetable in the lands where it is grown.

Habitat: Since it is a tropical marsh plant the dasheen needs a warm, humid atmosphere. It can never be too warm for it in our climate. It must be kept under glass all year round, therefore the living room or garden is not suitable. During a particularly hot summer, however, it may be placed in a warm, sheltered spot, for example against a sunny house wall. Temperatures of 25 to 30°C suit the plant best.

In autumn the foliage dies back. The pot containing the corms is kept dry until the following spring at temperatures which should not be allowed to fall below 15°C.

To allow its large leaves to develop fully the plant needs plenty of space. A fully grown dasheen with its huge leaves is a splendid and most decorative plant.

Soil: Originally a marsh plant, *Colocasia* needs a very damp, indeed, wet, soil, which should not contain peat. Sandy loam has proved a good medium. The dasheen is one of the few useful plants which tolerate lengthy periods of waterlogging.

Feeding: At the onset of growth an organic feed dissolved in water can be given weekly. In order to obtain the largest possible leaves, good feeding is necessary in addition to warmth and light. By dissolving organic dung in water and pouring it into the pot, we can practically allow the plant to stand in its own nutriment. Cease feeding in August to allow the corms to mature.

Maturing, harvesting: *Colocasia esculenta* is cultivated for its corms. They are boiled like potatoes, but since they contain calcium oxalate crystals the water should be changed once. The corms are rich in carbohydrates. Young leaves and their stalks can be eaten as a vegetable. It is, however, advisable to add a little soda in order to neutralize the calcium-oxalate crystals.

Propagation, cultivation: Daughter corms from a fully developed plant supply us with new plant material. They can only be obtained from countries where the *Colocasia* grows or from botanical gardens. In spring plant them in a pot. The pot, or preferably tub, should not have a drainage hole. Plant the corms several centimetres below the soil surface and place in a warm, light position. The potting material can be kept very damp from the start. After growth has started the water level should not quite reach the corms.

From the onset of growth till the leaves begin to turn yellow, the plant can be cultivated in water. It is equally possible to grow it in a pot or tub in potting material which is constantly wet. In the long run it will not tolerate tap water.

Turmeric
Curcuma longa
Family: Zingiberaceae

Dried and ground turmeric corm is an ingredient of curry powder and also the reason for its yellow colour.

The colouring substance in the rhizome is used industrially for the colouring of food and it is also used in the textile and leather industries.

Origin: It is a native of South Asian regions. Today it is cultivated in India, China and the tropical regions of South East Asia.

Habit: Long-stalked herbaceous leaves grow out of a knotty rhizome. They are about 40 cm long and 10 cm wide and of a rich green colour. The veins are very marked. The plant attains a height of ½ m. In our climate the aerial parts die off in autumn and the rhizomes put out fresh growth in spring.

Habitat: Turmeric should be grown under glass, although it may be kept in a sunny spot in the open air in the warmest part of the year. It is not suitable for the living room as it requires high humidity. In dry conditions the plant is very susceptible to red spider mite. Full sunlight is tolerated, but under glass the plant needs shading slightly during the hottest days, otherwise the herbaceous soft leaves will get scorch marks. After the rhizome has matured in the autumn, the leaves die off. The tubers must be kept dry in the plant pot and overwinter at a temperature no lower than 12°C.

Soil: Turmeric demands a well-aerated, humus soil which drains well. Like all useful plants which have rhizomes, it will not tolerate prolonged wetness or waterlogging (the exception is *Colocasia*

esculenta, cf. p. 54). The plant will grow happily in standard potting material. The rhizomes will go prematurely woody in heavy loamy soil.

Watering: Turmeric requires an even, slight moisture in the soil ball from spring until into autumn. From October give only enough water to prevent the soil from completely drying out. The condition can be gauged by feeling the weight of the plant in its pot. When the yellowed leaves are removed the soil must be left to dry out until the time for transplanting arrives in the spring. The plant will not tolerate tap water in the long run. Use rain-water or softened water.

Feeding: Turmeric appreciates an organic feed every 4 weeks. Chemical fertilizers may be used, since the soil is renewed every year and mineral accumulations therefore do not occur.

Maturing, harvesting: Turmeric is a sharp, piquant spice similar to ginger. The rhizomes are lifted, cleaned and boiled. They are then dried and ground. The attractive yellow colour is acquired during the cooking process. Turmeric gives curry powder its characteristic colour. It may also be bought on its own, as a single spice. Curry powder consists of at least 9 ingredients: turmeric, ginger, cardamon, coriander, caraway, nutmeg, pepper and cinnamon.

Propagation, cultivation: Turmeric is propagated by means of the rhizomes or parts of the rhizomes. It is only possible at the moment to obtain them in their native countries or from botanical gardens. Cover lightly with soil and keep light and warm between 20 and 25°C. They should be planted in early spring and the first shoots will be visible after 3 to 4 weeks.

Tiger nut, chufa
Cyperus esculentus
Family: Cyperaceae

A good cooking oil is pressed from tiger nuts in Mediterranean countries. In Spain a milky drink, known as Horchata, is prepared from the boiled tubers; flavours and spices are added and it is drunk or made into ice cream. Tiger nuts are very rarely available over here.

Origin: The plant was known to the ancient Egyptians. Today it is cultivated in the southern Mediterranean regions, in West Africa, south-western Africa, India and Brazil.

Habit: It is a sedge. Tubers the size of an acorn develop on thin subterranean stolons. The plant produces no seed. The leaves are grasslike, 20 to 40 cm long.

Habitat: The plant is an annual. It may be placed in a light south facing window or in a sheltered spot in the open air. Shelter from the rain is only necessary in damp summers. Overwintering is not a consideration, since the tiger nut is grown from scratch every year.

Soil: It requires a light humus soil mixed with sand, which should tend towards the acidic.

Watering: It should be watered with rain-water as it will not tolerate tap water for any length of time. In areas where the water is not particularly hard, however, tap water can be used. Prolonged wetness will kill the plant whereas short periods of dryness are an advantage. Waterlogging leads to root rot.

Feeding: Since several specimens are planted close together in the flower pot, a good deal of nutrition is required, so we must feed every fortnight. A chemical fertilizer may be used as the plant is not very sensitive to mineral salts in the soil.

Maturing, harvesting: The harvesting of the tubers takes place in October/November. For this purpose, remove the entire tussock from the pot. The tubers are ripe when their skins are brown. They are separated from the grass and kept dark, cool and dry and used for propagation the following year. The aerial parts may be put on the compost heap.

Propagation, cultivation: The tubers lifted in autumn are planted fairly close together in the pot in which they will remain throughout the growth period. They cannot be transplanted once they have started to germinate. The first shoots appear after about 3 to 4 weeks, and the growth period in our climate lasts from 6 to 8 months.

Below left: Tree
tomatoes are
similar in taste to
our usual
tomatoes.

Below right: The
tubers of *Dioscorea*
formed in the leaf
axils can weigh up
to one kilogram.

Tree tomato

Cyphomandra betacea

Family: Solanaceae

Seeds of the tree tomato are available from gardening firms all year round.

Origin: Northern South America.

Habit: The tree tomato will attain a height of about 2 m in our climate. It has large, heart-shaped dark green leaves. In the first few years the plant has the appearance of a green herb; only later will it become woody.

After about 3 years the first flowers appear. They are pinkish white and are arranged in a pendulous panicle. From them develop fruits which resemble hen eggs in size and shape. The mature fruit are a deep purplish red in colour.

All parts of the plant have an unpleasant smell, but this is only evident if you touch or break off part of the plant.

Habitat: It is suitable for the living room or for the greenhouse. Although the plant is not too delicate, it should not be put outside as

rough weather will easily damage its large leaves. It is better to plant it in a pot than in a border, otherwise it will grow too big. It should be protected from the sun: a semi-shaded position is best. The large leaves are very sensitive to sunlight if the plant is kept under glass. In winter the plant may be kept in a heated room. A light cool room will also do if the water given is correspondingly reduced.

Soil: The tree tomato is not very fussy about the type of soil. Standard potting material is adequate.

Watering: Water evaporates from the leaves in quantities dependent on the temperature. It is sometimes necessary, therefore, to water both morning and evening. Depending on the room temperature, only a slight moisture in the soil should be maintained in winter.

Feeding: A fourteen day cycle is enough since the plant will reduce flowering and become luxuriant if the nitrogen supply is high.

Maturing, harvesting: When the red fruits yield to the touch, they can be picked. They may be eaten raw or used for cooking in the same manner as ordinary tomatoes. They taste similar.

Propagation, cultivation: The seeds extracted from the fruit should be laid on a humus soil and covered lightly. Covered with polythene film the pot should be put into a shaded position at about 20°C. Germination will take place between 14 days and 3 weeks.

There should be only a slight moisture in the pot. As soon as the cotelydons have uncurled the film may be removed.

Potato yam, air potato
Dioscorea bulbifera
Family: Dioscoreaceae

Tubers of this plant are sometimes obtainable in shops catering for ethnic minorities.

Origin: Hundreds of species are native to all tropical and to parts of the subtropical regions of the world. About ten of them are cultivated for human consumption.

Habit: All *Dioscorea* species are climbers. They have a herbaceous stem several metres long. The leaves are large, usually opposite, heart-shaped and sometimes prettily patterned. The flowers are small and appear in loose racemes. They are monoecious or dioecious.

Tubers up to 1 kg in weight are formed in the axils of the petioles. They look like potatoes. The useful part is the root-tuber formed underground which can, in good conditions, be over 50 cm long and weigh up to 20 kg.

Habitat: Potato yams can only be grown under glass at temperatures over 20°C. An exception to this rule is *Dioscorea batatas*, which will thrive at temperatures between 15 and 20°C. *Dioscorea bulbifera* described here needs much sunlight: only put plants in the shade at the hottest time of year. Air humidity should be high, values over 80% doing no harm. Potato yams should not be put outside in the summer months. Cultivation in the living room is not possible. In specially adapted windows which afford good growing conditions the plant growth becomes too extensive.

In late autumn the leaves turn yellow, the aerial parts die off and may be removed. The tubers should be left in the plant pot undisturbed, dry and not too cold. Replanted in spring they will send out new growth.

Soil: The yam root needs a well aerated humus soil with loam content. A heavy potting medium which tends to become a solid mass is unsuitable, since water retention will lead to immediate root rot. Standard potting soil with a ⅕th addition of lime-free sand or perlite has proved successful.

Watering: The potato yam needs an adequate amount of moisture but on no account wetness. The plants make full use of any moisture in the soil and will continue to grow when other plants would wilt. They are sensitive to prolonged wetness and the soil should almost dry out between waterings. The water content of the soil may be estimated by lifting the pot and judging from the weight. Water with lime content will not be tolerated long term.

Feeding: Since the yam forms a great mass of leaves in a short time, it needs an organic feed fortnightly from the onset of growth until the end of August. As recommended for all other plants described here which are grown in pot or tub, organic feeding is desirable in order to avoid the build-up of mineral salts in the soil caused by the chemical type. There are remarkably few cases of deprivation damage or other disadvantage as a result of this policy.

Maturing, harvesting: The tubers are ready for harvesting after 9 to 11 months. They are lifted and used like potatoes. In Africa a purée made from boiled yams forms part of the staple diet. Chips may also be made. The bulbils which are formed in the leaf axils are not consumed.

Propagation, cultivation: Potato yams are propagated by cuttings or by using the smaller sowing tubers. The axillary bulbils may also be used. When the main tuber gets too big, the upper part may be cut off and replanted after the cut has dried.

The tuber can be encouraged to put out growth as early as mid-February in order to give the yam the longest possible growth period in our climate. The tubers are given only a thin covering of soil in their container, watered slightly and placed immediately in a warm light position. Growth starts after 3 to 5 weeks. Each plant needs a stick or line down from the greenhouse roof as a climbing support. The plant will attach itself to this and

The flesh of the persimon is juicy, sweet and has the texture of the tomato.

grow up it.

Tubers from the previous year are lifted from the old soil, dead roots are removed, and the tubers replanted in fresh potting mixture. Kept light and warm they will put out shoots more quickly than the axillary bulbils.

Persimon
Diospyros kaki
Family: Ebenaceae

The fruits are obtainable in autumn and winter in most well-stocked fruit shops.

Origin: A native of China and Japan, it is now grown throughout the world in sub-tropical zones.

Habit: The persimon tree attains the height of an apple tree in the countries where is is grown. It has large ovate-lanceolate leaves with a surface like tanned leather. The large yellowish-white flowers are situated in the leaf axils and open in spring. They are self-pollinating.

Habitat: *Diospyros* should be put outside after the May frosts in a sunny place; if possible sheltered from the wind. There it will stay until the return of the first frosts. Since the plant sheds its leaves in autumn, a moderately light place is adequate in winter. The temperature should not exceed 5°C. Since it is adapted to dry air, the plant may also be kept in a light living room window. In southern parts of England it can be grown in very sheltered sunny positions.

Soil: A potting material somewhere between peaty soil and a cactus medium has proved successful; that is, a soil which is not too

nutritious and which drains easily.
Watering: The persimon is fairly resistant to dry conditions and watering should be done with caution. Water should not be given until the pot or tub is almost dry, but then give copiously. Rainwater is best. Since the tree is leafless in winter the soil ball should be kept only moist.

Feeding: As long as the tree has not reached the flowering stage a feed every four weeks is all that is necessary. When it starts to form fruit it should be increased to every two weeks. Cease feeding at the end of August.

Maturing, harvesting: The fruits on sale here are all seedless and therefore useless as a source of new stock. The tree, however, is available from some nurseries.

Oil palm
Elaeis guineensis
Family: Palmae

Fruits (seeds) of this plant are not usually obtainable here. Certain

The oil palm needs a bright position in the greenhouse. Only in very hot weather conditions can it be allowed outdoors.

horticultural firms which specialize in tropical plants occasionally sell them.

Origin: There is, as yet, no certain scientific proof as to where the plant originated. Some botanists assume it to have come from the tropical regions of South America, others similar regions in Africa. Today it has a worldwide distribution within a belt between 10° latitude either side of the equator.

Habit: The oil palm has a slender stem which can attain 15 to 20 m in greenhouse specimens and up to 30 m in nature. Pinnate leaves form the crown. The leaf basis remains attached to the trunk several years after the leaf has fallen, then is itself shed. Only then does the trunk gain a further smooth section. It can reach a diameter of up to 60 cm.

Habitat: The oil palm needs a light position in a warm greenhouse all the year round. It cannot get too warm for it there in summer. In winter the temperature should not fall below 18°C. A tropical palm, it thrives best in a hot damp atmosphere. It is not possible to cultivate it in the living room and even in our hottest summer it will not do well out of doors. Nevertheless, kept under glass it needs shading slightly in the summer months or the leaves will scorch. The average temperature should be around 25°C. High air humidity, it must be emphasized, is vital for this plant.

Soil: The plant needs a loose humus soil. Standard potting material is suitable. The soil should be neutral to slightly acid. A large peat content gives the leaves an intensive deep-green colour.

Watering: *Elaeis guineensis* needs an evenly moist soil ball. It will react negatively to alternating dry and wet conditions. If the winter temperatures reach its lowest point of tolerance, cut down on water. An almost dry soil ball at that time of year will not hurt the plant. It does not like water with lime content very much, so here again rainwater is best. It is very important to give water at room temperature in winter.

Feeding: A feed every 4 to 6 weeks during the growth period should ensure good development. Cease feeding from the beginning of September. The oil palm does best on a mixture of manure with a small amount of chemical fertilizer to guarantee its supply of trace elements.

Maturing, harvesting: In the conditions the amateur can provide for it, it will not bear fruit.

Propagation, cultivation: It is propagated by means of its seed. If you have acquired some fruits, remove the seeds from them. Soak them for up to 36 hours in warm water at temperature of 30 to 35°C. This procedure is necessary as they do not germinate easily. The seeds can also be placed in a 1% solution of hydrochloric acid for 2 days. Wash the seeds before sowing.

Sow them 1 cm deep in standard

A ten year old
loquat, even
when grown in a
tub, may yield up
to 35 kg of fruit.

potting mixture, laying the seeds horizontally and not pointing upward. A piece of film placed over them will retain the air humidity. Fresh seeds will germinate after about 12 weeks, but a temperature of 20 to 25°C must be maintained. Even if you receive seeds in the winter, you must sow at once, since they quickly lose their viability. Older seeds may take more than a year before they germinate.

When the pointed cotyledons appear, continue to wait until up to 3 leaves are there. First of all, 3 to 4 pointed leaves will develop, then some swallow-tail shaped leaves, then the plant starts to produce its well known pinnate leaves.

Transplant with the utmost care, so that the seed still connected to the plant does not come adrift. If this happens, the plant will die, since the nutrients pass through the seed.

Loquat
Eriobotrya japonica
Family: Rosaceae

The fruits, known as loquats, are sometimes for sale.

Origin: The loquat is a native of China and Japan. Today it is cultivated at higher altitudes of the tropics and in all subtropical regions.

Habit: It is a small evergreen tree with a short trunk. In the areas of cultivation it grows 5 to 10 m tall.

The leaves are large, up to 20 cm long, 6 cm wide and lanceolate. The leaf margin is coarsely dentate, the upper surface dark green, the underside white and fuzzy. New shoots are a velvety silver-white. It flowers in autumn. The flowers are terminal, small, whitish and racemose. By spring plum-sized, pear-shaped light yellow fruits have developed.

Habitat: After the May frosts until the onset of the first autumn frosts the loquat may be kept outdoors in a warm and sunny position. It should be protected from longer rain spells. In summer it cannot be too hot for it; in winter a place in a cool room at 5 to 10°C is suitable. It makes a good indoor plant for a light, south-facing window. In the British Isles it can be grown outdoors in sheltered positions in southern counties, but it will not produce fruits.

Soil: The tree is not very fussy. A loose, well drained and humus-rich soil is acceptable. A standard potting mixture with a small addition of sharp sand has proved successful.

Watering: The loquat is adapted to dry conditions, as the fuzzy felty surface of its leaves shows. The soil may be allowed to get almost dry between waterings with no damage to the plant. Careful observation is needed, however, for if the soil becomes completely dry, the plant will shed its leaves. The tree requires a slightly acid soil, there-

fore it will not tolerate most tap water over a long period. In winter the soil ball should be just moist.
Feeding: To avoid excessive growth, the loquat only needs to be fed every six weeks. Cease feeding from September. The plant reacts extremely well to organic feeding.
Maturing, harvesting: The trees begin to flower when they are 4 to 5 years old. As they get older the crop is so thick that it becomes necessary to thin out the fruit. A 10 year old tree grown in a tub can easily yield 35 kg of fruit. In our climate the harvest time is from May to June. The fruits are picked when fully ripe. They can be eaten raw or made into a compôte. Commercially they are used for marmalade and jam.
Propagation, cultivation: The amateur grower will derive his crop from seeds. Commercial garden firms use grafting. The seeds are sown 2 cm deep in spring and the container placed in a light situation at a temperature of 20°C. Germination ensues in 4 to 6 weeks. During germination the small plants

Only the common
or Adriatic fig
should be
considered for
cultivation since
it will set fruit
without
pollination.

should not be placed in the sun, as the first leaves are very sensitive. Some tree nurseries offer *Eriobotrya* for sale as an ornamental garden plant.

Fig
Ficus carica
Family: Moraceae

The fruits are obtainable fresh in the summer months and dried the whole year round.

Origin: The fig is a native of Mediterranean countries.

Habit: The fig is a deciduous plant, and according to the way it is pruned, a shrub or tree. It can reach a height of 10 m. The large, dark green leaves are palmately lobed. Inflorescences are formed three times annually in the axils of the leaves. Kept as a tub plant in our climate the edible fig flowers once a year and produces sweet fruits in the autumn.

Smyrna figs are not suitable for our purposes on account of their complicated fertilization process, so only figs of the Adriatic type will be considered, since they are parthenocarpous i.e. fruits are formed without the fertilization of the ovary. Young figs which do not reach maturity in the year they are formed will continue to grow the following year.

Habitat: The fig prefers a sunny, not too damp spot. In rainy years it will appreciate a position where the soil ball can dry out. It will never be too hot for it outside. Slight shade should be provided during the hottest hours of the day. Since it loses its leaves in winter, a cool place, which does not need to be too light, is suitable. In sheltered positions and in southern parts of the British Isles it thrives outdoors. As a matter of fact many splendid fig trees can be seen in southern counties positioned against sunny walls. *Ficus* will yield a fine crop of mature figs (e.g. in Great Dixter).

Soil: The fig likes a well-drained, lime soil, but it will thrive in any soil which is not too rich in humus and peat. Excess water must drain off without causing the potting material to set into a mud. Standard potting material with ⅓ gravel suits the plant.

Watering: The plant does not need to be watered until the leaves begin to droop. Then water well until the soil ball is moist. Do not water again until the leaves begin to droop once more.

Feeding: An organic feed may be given every 8 days.

Maturing, harvesting: Well fed plants in a normal summer will provide a crop ready for harvesting in mid-August. The figs will not ripen simultaneously, but every 1 or 2 days a certain number will be ready so that fresh figs are available for weeks.

They must be picked immediately when ripe or fermentation will set in. Fresh figs will only keep for a

Of all the fruits yielded from *Citrus* and its relations only the kumquat (right) is eaten whole, including its peel.

The decorticated roots or stolons of *Glycyrrhiza glabra* (far right) are used for the production of all forms of liquorice.

short time, even when refrigerated. They are ripe when they feel soft and burst easily when pressed.

Propagation, cultivation: They are propagated by taking cuttings. These are taken from mature branches, preferably below a bud. They should be about 20 cm long and are planted with one third of their length beneath the soil-sand mixture. The pots containing the cuttings are kept shaded and at a temperature of approximately 20°C. See that the cutting has not been planted upside down, that is with the buds facing downward. If well cared for, spring cuttings will bear the first figs as early as the following autumn. *Ficus carica* is available from many nurseries in Britain.

Kumquat
Fortunella margarita
Family: Rutaceae

Kumquats are obtainable in fruit shops and delicatessens from autumn to spring.

Origin: The kumquat is a native of China and Japan.

Habit: It is a many-branched bush with leaves similar to those of the mandarin. It has a somewhat arching habit. Grown in a tub it can attain a height of more than 2 m in our climate. The small, white, scented flowers sit in the leaf axils of the lateral shoots.

After pollination with its own pollen has taken place, damson-sized orange-coloured fruits develop. These are the only fruits in the entire *Citrus* family which are eaten whole, including the peel.

Habitat: In accordance with their origin, kumquats are not particularly sensitive to cold; consequently after the May frosts they may be placed in a sunny spot in the garden until the next frost. They will fruit there without trouble.

Since they usually have a heavy fruit crop they should be protected from wind to avoid being dashed and broken. They should be kept light in winter, but not warmer than 5°C.

Soil: Any humus soil that drains well will suit *Fortunella margarita*. It is important not to turn the soil into mud when watering as this would endanger the oxygen supply.

Watering: Again, use rain-water and ensure an even moisture in the soil. Give only a limited amount of water in winter to avoid root damage.

Feeding: The kumquat will respond well to feeding. To obtain a rich fruit crop it should be fed weekly. Do not cease feeding until the end of August to permit full development of the wood. If the crop has been particularly heavy the plant will produce fewer flowers the following year.

Maturing, harvesting: As soon as the fruits are deep orange in colour

and yield when pressed with the finger, they are ripe. They can be used in many ways, e.g. as a garnish. Completely ripe kumquats are so sweet that they can be eaten raw as a dessert.

Propagation, cultivation: cf. *Citrus madurensis*, p. 44/45.

Liquorice
Glycyrrhiza glabra
Family: Leguminosae

Liquorice, on sale in its many forms, is obtained from the root of *Glycyrrhiza glabra*. It is an ingredient of many medicines, cough mixtures for example. It is also used to give aroma to tobacco.

Origin: It is a native of the Mediterranean region, but today is also grown in parts of Asia and is found in Central European wine growing areas.

Habit: A powerful root, yellow on the inside, bears procumbent thin stems with penny-sized pinnate leaves. Racemose inflorences with violet or creamy yellow butterfly-like flowers grow from the leaf axils. They develop into 3–5-seeded pods. The plant may be grown under glass but also out of doors; it must have a light, sunny position. Dry air will not harm it. It can continue to grow through winter, but it is better to give it a rest by reducing the watering and placing it in a cooler position. It will not be too warm in our climate in summer, and in winter temperatures of between 5 and 10°C will not harm it.

Soil: The plant demands a non-rich

soil, such as cactus potting mixture which is sold in the trade. It is important for superfluous water to drain away. Accumulated water in the soil would rot the roots.

Watering: Liquorice is adapted to a drier climate, therefore the soil may be allowed to dry out between the waterings. If the plant is kept in the open air it should be protected from prolonged rain. In winter the aerial parts die off according to temperature conditions, and in this case the soil ball must be kept only slightly moist until the onset of new growth. The plant will not tolerate tap water for long, so rainwater should be given.

Feeding: No more than occasional feeding is necessary. Make sure that you stop feeding after August, even if the plant is kept under glass, to enable the full development of the rhizomes.

Maturing, harvesting: The product derived from the rhizome of *Glycyrrhiza glabra*, a sticky, blackish-brown juice, is extracted by an industrial process. The confectionery industry and pharmaceutical firms then make use of it. It is used to give aroma to tobacco and to flavour some types of beer.

Propagation, cultivation: It is best to grow it from seed. The seeds are sown in cactus potting mixture in spring and covered only slightly with soil. Water lightly and keep in a light place at temperatures around 20°C, and the seeds will germinate, according to age, in 4 to 6 weeks. The plants may also be propagated by splitting. This should also be done in spring. Sometimes gardening catalogues have liquorice plants for sale.

Cotton
Gossypium herbaceum and *G. hirsutum*
Family: Malvaceae

Cotton seeds are regularly on sale in specialist shops and garden centres and they germinate easily.

Origin: The original birthplace of the cotton plant is southern Africa. Some scientists believe the American species come from the Andes region of South America.

Habit: The plants cultivated today are all shrubs about 2 m high. Branching occurs very early in the young plant. The cotton plant has 3–7-lobed dark green leaves.

The cotton plant – illustrated is *Gossypium hirsutum* – tolerates dry air and is therefore ideally suited for the living room.

The attractive flowers are short stalked and grow from the leaf axils. They have the typical maple-flower shape. The petals are white, light yellow to red. *Gossypium hirsutum* has pure white flowers, those of *Gossypium herbaceum*, yellow with a red honey guide. Both of these species are available to the amateur grower.

After successful pollination an ovoid capsule the size of a table-tennis ball develops. In *G. hirsutum* it bursts after 3 months releasing the seed fibres. In *G. herbaceum* the capsule remains closed even when ripe. The cotton plant can be grown as an annual or perennial.

Habitat: The plant needs a great deal of warmth: it can never be too warm for it in our climate. Temperatures between 20 to 30°C in summer are easily tolerated. In winter they should be no lower than 15°C. A light situation all year round is absolutely essential. It makes a good indoor plant, since dry air suits it. In the summer months it can also stand in a sheltered spot in the open air but must be protected against prolonged rain.

Soil: The cotton plant requires a light, loose humus soil mixed with sand. It should be slightly alkaline; acid soil is unsuitable. Superfluous water should drain away easily.

Watering: The plant tolerates regular watering during the period of leaf development. When flower development starts there should be

Right: Kenaf
displays a fine
flower. For the
amateur grower it
is important to
collect the seeds
for a new season.

only a moderate amount of moisture in the soil. Shorter dry periods do no harm. In winter the soil ball should be only very slightly moist if the plant is being grown as a perennial.

Feeding: Organic fertilizer mixed with a handful of chemical nutrients and dissolved in water will give good results. Feeding exclusively with chemical fertilizer can lead to damage, as the plant is sensitive to mineral salts such as sodium carbonate. In plantations where watering has to be carried out, the plant will tolerate up to 0.5% of common salt (sodium chloride) in the soil. Plants grown in pots should be fed every 14 days from March to September. If planted out, in which case it will show luxuriant growth, it needs feeding every 4 weeks.

Maturing, harvesting: In autumn when the capsules are ripe, the seeds with their seed fibres must be extracted. If you take the trouble to separate the fibres from their seeds, you would have a sufficient cotton wool to polish a car.

Propagation, cultivation: The cotton seeds should be sown in spring 1 cm deep in the seed bed. The soil temperature must be at least 18°C. With sufficient moisture and warmth the seeds will germinate in a few days. After the first 2 to 4 regular leaves have appeared they may be transplanted, taking care not to damage the root system. Sowing should take place as early as mid February as a later sowing might not result in flowers.

Kenaf
Hibiscus cannabinus
Family: Malvaceae

Since the fibres are only of industrial importance, no plants are found in horticultural catalogues.

Origin: The kenaf is from the Afro-Asian region. It is grown in those areas of the tropics and subtropics where jute will not grow.

Habit: It is a coarse-stemmed shrub, up to 4 m in height, with lobed leaves. The attractive flowers bloom singly in the leaf axils. They are up to 10 cm in diameter and are pale yellow with a dark red to dark brown patch in the centre. After self-pollination the capsule develops.

Habitat: The plant needs a warm light place under glass. It can only be put outside at the height of summer and then in a sheltered spot. Direct sunlight is not harmful even if under glass.

Soil: It is not particularly fussy in the matter of soil. Standard potting material is entirely suitable.

Watering: The plant demands a great deal of moisture in the soil, but water must not stagnate there. Tap water will not be tolerated over a long period of time. If the plant is too dry, the flowers wilt, but the plant will recover after the next watering without any damage.

Below: Sweet
potatoes have the
taste of potatoes
which have been
exposed to frost.

it ripens becomes brown. When it begins to burst open, the seeds are ripe. They are small and round and the size of a garden pea. They must be kept cool and dark during winter. Sown on a peat seed bed in February, they will germinate at temperatures between 20 and 25°C within 4 weeks. From germination to flowering takes 6 months. The seeds are then ripe after a further 2 months.

Sweet potato
Ipomoea batatas
Family: Convolvulaceae

Obtainable in fruit and vegetable stores in the summer.

Origin: The sweet potato probably comes from South America, and its name is of South American Indian origin. It was cultivated by the

Feeding: As soon as the plant has developed its fourth leaf, it may be fed on a weekly basis with an organic fertilizer until the seeds are ripe.

Maturing, harvesting: The amateur will only be interested in obtaining seeds, since the plant's main product are the fibres contained in the stem. These must be extracted by an industrial process. They are considerably finer than jute fibres and can therefore be used to make finer textiles than sacking. After the seeds have been harvested, the plants may be thrown on to the compost heap, since annual cultivation presents fewer problems than overwintering.

Propagation, cultivation: Since the flower is only open for a day it must be pollinated immediately. The fruit capsule is green at first and as

The sweet potato is an annual plant with fresh-green lobed leaves, heart-shaped in outline. It will only flower in the tropics.

natives in pre-Columbian times and was introduced into Europe before the ordinary potato. Today it is grown throughout the world in the tropics and subtropics.

Habit: The sweet potato is an annual plant forming creeping stems on the ground with opposite, heart-shaped leaves. These leaves are approximately 5 cm long, equally wide and a fresh green colour. The flowers, which will only be formed in the tropics (short-day plants) are funnel shaped, white or reddish, and resemble those of the bindweed. The stem can be several metres long in a damp climate. The tubers are spindle-shaped, up to 20 cm long, yellow to reddish in colour and up to 3 kg in weight.

Habitat: The sweet potato needs temperatures between 25 and 30°C and direct sunlight. it can only be grown in the greenhouse in our climate or in a specially designed tropical display window. It is impossible to grow it in the living room or garden. High air humidity of 80% is necessary, otherwise the plant will be infested by red spider mite. Night temperatures should not fall below 12°C, since lower temperatures will cause the plant to die off. It is a typical sun loving plant and only in the midday hours of the hottest season do we need to provide some shading.

The foliage begins to die off in autumn and the plant should be moved to a cooler position (not below 15°C). The tubers are left in the pot and the soil kept slightly moist. Placed in a light, warm spot from mid-February, it will begin to put out new growth.

Soil: cf, p. 10.

The bay tree displays many axillary yellowish white flowers in Spring.

like normal potatoes, but have a slightly slippery texture and are sweetish, rather like potatoes which have been exposed to frost.

Propagation, cultivation: The sweet potato is propagated vegetatively, either by means of stem cuttings or the new tubers formed during the growing period. 20 to 30 cm long pieces of the plant with 2 to 3 buds – the whole stem is suitable for this – are set in peaty soil. The cuttings may also be laid flat on the surface and covered lightly with soil. Kept warm and light and covered with polythene film, they will take in 8 to 14 days. As soon as the new growth is visible, the film covering must be removed.

Bay tree
Laurus nobilis
Family: Lauraceae

The plants are sold in specialist nurseries and garden centres.

Origin: It has a wide distribution in the Mediterranean region, especially in Corsica.

Habit: *Laurus nobilis* is a characteristic part of Mediterranean vegetation. The sclerophyllous plant can attain a height of 10 to 12 m. The leaves are leathery, shiny and have entire margins on the upper side, c. 10 cm long and up to 4 cm wide. The leaves are kept all year round and the individual leaf will live for 2 to 3 years. In spring it blooms, with

Watering: To start with, the sweet potato will need a good amount of water as it develops a huge leaf mass in a short time. Water containing lime will not be tolerated long term. It is also very sensitive to water that is too cold, and will react with tuber rot. The water quantity should be reduced from July–August to encourage tuber development. In late summer the plant may be allowed to dry out for short periods between watering.

Feeding: Since the plant grows very quickly, it should be fed every 8 to 14 days from March to August. Cease feeding in autumn to allow the tubers to mature.

Maturing, harvesting: When the foliage dies off in late autumn the tubers may be lifted. Once out of the ground they do not keep long. Rich in carbohydrate, they are used

Given proper care
Laurus nobilis
will develop into
a beautiful tree
with dark green
foliage.

many small yellow-white flowers appearing from the leaf axils. The tree is generally grown in a tub, and if not pruned will turn into a beautiful dark green-leaved specimen up to 2 m or more high.

Habitat: It should not be kept in the living room, nor under glass, but in an outdoor spot where it receives sufficient sun. It can survive in semi-shade, but will not develop its characteristic form and thick foliage. It does not tolerate frost, but as soon as the temperature rises to a few degrees above freezing, it should be put outside. There it should stay until the onset of autumn frosts. In winter it should have a frost free position which does not have to be particularly light. The temperatures should not be higher than 5°C. In southern counties of England *Laurus nobilis* is fairly hardy.

Soil: This should be slightly acid. Standard potting mixture with a third of peat will do. Do not repot until pot-bound.

Watering: In the long term the bay tree will only tolerate lime free water e.g. rain-water. It does best when the soil ball is evenly moist, less so in winter. The plant needs watering in the winter season too, since it retains its foliage. If the soil ball dries out, the plant will take a long time to recover.

Feeding: An organic feed in the vegetation period every four to six weeks is enough for our bay tree, otherwise it will grow too excessively for winter storage. It may be cut back in spring.

Maturing, harvesting: The bay tree's aromatic leaves are used in perfumery and cooking. They can be picked and used fresh from the tree all year round. They are much more aromatic than the dried ones in the shops and are particularly favoured for the seasoning of game dishes.

Propagation, cultivation: If you do not wish to buy an established plant as such, but wish to grow your own, you must acquire some cuttings. These are not, in fact, cut from the tree, but torn off with a small strip of the branch bark, if possible. This 10 to 15 cm long cutting is set in peat, watered and placed in a shady spot at c. 15 to 20°C. Planted in spring, they should show root development after a few weeks.

Litchi
Litchi chinensis
Family: Sapindaceae

The fruit is available, tinned, all the year round. The fresh fruit is available from spring to early summer.

Origin: The litchi comes from southern China but is now grown throughout the drier tropical zones of the world.

Habit: In lands where it is cultivated it grows to a height of 10 to 15 m. It has narrow, oblong, pinnate leaves which taper to a point.

The flowers of
Luffa aegyptiaca
(right) are similar
to those of our
climbing
greenhouse

cucumber. After
maceration the
remaining
vascular bundles
form the loofah of
commerce.

The inflorescences bear up to 30 white pentamerous flowers. They are male, female or bisexual. The relationship between the flowers differs from year to year. The fruits are the size of large cherries and have a perfumed white flesh under their granulate leathery skins.

Habitat: After the summer growth period the plant requires a definitely cooler and drier resting time. In summer it may be placed outdoors and will not be harmed by direct sunlight. When grown under glass it needs shading slightly in the hottest months or the leaves will be scorched. In winter it should have a light but cooler position. Care must be taken that it does not start to grow too early in spring.

Soil: The soil for *Litchi chinensis* should not be too humus rich. Up to 50% sand and loam content will ensure good drainage and an even moisture in the potting mixture. This moisture content must be maintained in winter, but avoid wetness.

Watering: Its water requirements are fairly high during the growing period. Since the plant reacts badly to lime content only clean rainwater or treated tap water should be used. In spite of its water needs, the surface of the soil should be allowed to get dry from time to time, since waterlogging would not be tolerated. Irregular watering, alternating dry and wet conditions, will result in leaf fall. If the leaf tips go brown, this shows that mistakes with the watering have been made.

Feeding: Give the litchi an organic feed every 4 weeks in summer. This is sufficient since the plants do not get big enough in our climate to bear flowers.

Maturing, harvesting: The plants will not bear fruit in our climate.

Propagation, cultivation: In spring, sow the seeds extracted from fruits just purchased in a humus material about 1 cm deep. Keep slightly moist and in semi-shade at about 20–25°C. Depending on how fresh the fruit was, they will germinate in 2 to 3 weeks. Once this has happened and the first 4 leaves have been formed, they may be planted singly.

If you can obtain a plant, from, for example, a botanical garden, you could try to propagate with cuttings. The cutting should be woody, however, as shoots that are too green will not take, but go rotten.

Loofah
Luffa aegyptiaca
Family: Cucurbitaceae

Only the fruits of *Luffa aegyptiaca* are sold. The dried vascular bundles of the fruits are sold for use in the bath, or are used to make insoles. Young fruits of the non-bitter type may be eaten as a vegetable.

Origin: It is cultivated throughout the tropics.

Habit: The loofah plant is very similar to the well known climbing cucumber plant we see in greenhouses, with the same large lobed leaves. It also has yellow flowers. From the latter develop long cucumber-like fruit (botanically speaking berries) up to 50 cm long. The plant is annual and dies off in autumn.

Habitat: *Luffa aegyptiaca* is cultivated all year round under glass. It needs a light position but protected from the most brilliant sun. Temperatures around 25°C suit it. There is no point in attempting to grow the loofah outdoors in regions where a vine would not do well. It is possible to grow it in the living room, but the results will not be very good as, like all Cucurbitaceae, it thrives best when there is high air humidity.

Soil: The plant needs a rich, well drained and loose humus soil. I have grown my healthiest and strongest plants in a potting mixture with leaf litter.

Watering: The loofah requires an even moisture in the soil throughout the growing period. As it gets more extensive, more water must be given. Since the plant develops huge leaf masses, it may require two waterings daily if kept in a tub. Make sure that the water is the same temperature as the surroundings, as the plant will react to cold water by stopping growth.

Feeding: To obtain large fruits, the loofah should have organic feed of dissolved manure every 8 to 14 days. Feeding may be continued until mid September.

Maturing, harvesting: After fertilization of the flowers, the loofah fruit develops very quickly until it is up to 50 cm long and dark green in colour with light marbling. It remains attached to the plant until the foliage dies off. The fruits are then picked and kept dry and warm for a few weeks. When the outer skin has started to break up, it can be totally removed. Now we see the web of fibres from which the seeds must be extracted. The remaining fruit flesh can be removed by drying or washing and we are left with a usable loofah.

Propagation, cultivation: Fertile seeds, which can only be obtained in the tropics or from a botanic garden, are planted sideways in

small pots as early as the beginning of February. The quantity is usually two per pot. They are watered lightly and immediately placed in a warm, light position. To increase humidity, the pots may be covered with a layer of polythene. Fresh seeds germinate in a few days. Cultivation and care is as for the ordinary cucumber plant.

Mango
Mangifera indica
Family: Anacardiaceae

Mangoes are obtainable fresh in supermarkets and fruit shops the whole year round.

Origin: A native of the Indian subcontinent, the distribution of the mango now extends throughout the tropics and partly into subtropical regions.

Habit: The original mango tree grows up to 30 m tall but modern cultivars are much shorter. The mango tree has narrowly lanceolate green leaves, which when first formed are soft and reddish in colour, but change and become leathery.

The paniculate inflorescences develop at the tip of the previous year's shoots. The flowers are white or reddish in colour. The fruits are spherical, ovoid-globular or kidney-shaped and can weigh up to 1 kg. They contain a large, fibrous stone. The fruit flesh is delicious.

Habitat: In our climate mango can only be cultivated under glass where it will be light and sunny. Shade slightly at the hottest time of year. Since it develops an extensive root system, the plant container

Mangoes can be
obtained in the
shops virtually all
year round.

should not be too small. Planted free in the greenhouse the mango will develop this extensive root system and other plants may suffer. It should overwinter in light conditions at about 15°C. Lower temperatures will only be tolerated if the soil is dry.

Soil: The mango likes a nutritious and well aerated humus soil. It will quickly grow large in a sandy loam soil with a mixture of peat. As a result of its speedy growth and its extensive root system, annual transplanting is a necessity.

Watering: The aim must be a good supply of water but avoid accumulation in the soil. Evenly moist soil is important for the plant. Not until it is several years old and more resistant to drought, can we allow shorter dry periods without danger.

Feeding: Organic feeding has proved more successful than the chemical equivalent. A manure feed should be given every 14 days in the growth period. A dose of aluminium sulphate 2 to 3 times a year will encourage development. Cease feeding in autumn, since the mango will stop growing as the light intensity fades.

Maturing, harvesting: The plant will not flower or fruit in our climate, but is attractive for its leaves.

Propagation, cultivation: Remove the fibrous stone from a ripe mango fruit bought in summer and clean off the fruit flesh. The cleaning is essential to avoid committing hara-kiri with the knife when opening the stone, which is as slippery as a fish.

Unopened mango stones hardly ever germinate, since the fruit are still not ripe when harvested, and the seed does not often have the strength to burst open the hard shell. Open the stone at the narrow end with a sharp knife. Stick the knife point carefully into it and when it has penetrated a few millimetres, widen the slit with a twisting movement until you can force your finger in. Always begin to open the stone in the middle of the dorsal side so as not to damage the tiny seed.

The opened stone should now be set in humus soil with the seed downwards and the pot covered in polythene. It should then be kept in the shade at about 20 to 25°C. During germination the stone will hardly separate and will not lift out of the soil.

After about 3 weeks the first pinkish leaves will appear. From this point the polythene may be removed and the young plant placed in a lighter spot.

Cassava
Manihot esculenta
Family: Euphorbiaceae

Health food shops sell the product made from the starch contained in the tubers of this plant as tapioca,

which is used in puddings and some types of confectionery.

Origin: The plant is a native of South America, but had already reached Africa and Indonesia by the 16th century. Today it is grown throughout tropical and sub-tropical regions. It is unknown in its wild state.

Habit: The cassava is a short-lived shrub, which can be up to 3 m in height. The slightly zig-zag stem is filled with soft pith. The leaves are long-stalked and opposite. They are palmately divided with 2 to 7(9) leaflets.

The inflorescence is a terminal panicle. The male flowers are at the top of the inflorescence, the female ones below. Since the female flowers open about 8 days before the male, pollination from the same inflorescence is not possible.

The useful part of the plant is the root tuber, which is 30 to 50 cm long. It can weigh up to 5 kg. The leaves, boiled, can also be eaten as a vegetable. The eating of the tubers is not advised, for unless they are prepared by experts they contain the highly poisonous hydrocyanic acid.

Habitat: In our climate *Manihot esculenta* needs to be provided with a warm humid atmosphere with temperatures over 25°C during its growth period in the summer. Direct sunlight will not hurt it. It will not thrive in the living room or garden on account of its need for high air humidity, around 80%. Kept at temperatures between 15 and 20°C from November/December and with less watering it will mature and shed its leaves. It will survive the winter without difficulty in the above mentioned temperatures, given that the soil is kept slightly moist. It will put out new growth when it is moved to a light and warmer position from February onwards.

Soil: It demands a humus rich soil that drains well. Standard potting material with an admixture of sandy loam is best. If it is grown as a pot plant, a layer of sand several centimetres deep should be put at the bottom of the pot to avoid water retention, otherwise the tubers will rot very quickly.

Watering: An even moisture in the soil must be provided during the growth period. Sensitivity is required in the matter of watering, for if a sunless period follows after water has been given, tuber rot can quickly set in. As in the case of all perennial plants kept for long periods in the same container, rainwater is best if mineral salt accumulation is to be avoided. In winter when the shrub is leafless, watering with lukewarm water is essential. The tubers are sensitive to dousing with cold water and will suffer damage.

Feeding: The plant grows very quickly in spring, therefore a solution of organic fertilizer can be given every 14 days. From June–July the frequency can be reduced to a feed every 4 weeks. Cease feeding from September until spring.

Maturing, harvesting: When the leaves turn yellow or fall off, the tuber is ripe. In countries where it is cultivated the tubers are finely grated. Water is added to the resulting mash so that the starches sink to the bottom. The water used for this purpose then smells strongly of hydrocyanic acid. The starches are processed into flour or flakes.

Propagation, cultivation: *Manihot* can be propagated by means of the harvested tubers. Today, however, it is almost always propagated by means of cuttings. The cuttings are taken from the mature parts of the stem. The soft tip and the very woody lower stem are not suitable. The individual cutting should be about 15 cm long and possess several buds. The best angle is about 60°, for then the cutting will produce growth along its entire length. Vertically planted cuttings will only produce growth at the upper end. Kept at a temperature of about 20°C, in light conditions and slightly watered, they will begin to produce shoots within 2 weeks. *Manihot* can be propagated with cuttings throughout the year.

Sapotilla
Manilkara zapota
Family: Sapotaceae

Fruits are sometimes sold in fruit markets and in the fruit sections of supermarkets.

Origin: A native of the plains of northern South America, it is now cultivated in all tropical regions, especially in Central and South America.

Habit: The sapotilla tree grows to 25 m in its homeland. The large leaves, c. 10 cm long and 2 cm wide, are elliptic, leathery, shiny, and dark green. The flowers are small and white and situated in the leaf axils of terminal branches. The fruits are almost the size of apples with yellowish-brown transparent flesh. They can only be eaten ripe, since before this stage they contain much tannic acid and latex. When ripe, however, they are very sweet.

Habitat: The plant is strictly adapted to tropical conditions. It can therefore only be grown under glass in our climate. It is not possible to grow it in the living room or garden since it needs a constant air humidity of over 80%. Temperatures can rise above 30°C in summer and should not fall below 18°C in winter. It tolerates direct sunlight when not directly behind a pane of glass.

Soil: The soil should tend towards the acidic, be light and drain easily. The roots of this plant need a great deal of oxygen. It will not tolerate soil which retains too much water. The leathery leaves are also an indication of a preference for drier soil conditions. In winter the potting mixture should be only slightly moist. During this season the plant will not grow on account of the lack of light.

Watering: An even moisture must be maintained in the soil ball during the growth period. In winter it is kept almost dry. It does not matter if at this time the plant sheds a few leaves: the new growth will be even bushier in spring. It will not tolerate tap water.

Feeding: The tree should receive an organic feed every 4 weeks from March till the end of September. Organic fertilizers are again by far the best in my experience.

Maturing, harvesting: The sapotilla tree will only start fruiting after several years. Trees 5–6 years old are capable of flowering in our conditions. The fruit is almost the size of an apple; apically it has a navel-like depression below a persistent calyx. The sweet pulp is somewhat yellowish and translucent. Only when fully ripe is the fruit edible. It will keep for several days in the vegetable compartment of a refrigerator. The latex of the tree is exploited industrially: it is the basic ingredient of chewing gum, for which reason the plant is sometimes called the chewing gum tree. This latex is present in all parts of the plant.

Propagation, cultivation: The sapotilla may be grown from seed. In the lands where it is grown commercially the best cultivars are grafted on to the stock of the same species. Each fruit has 8 to 12 hard, shiny black seeds. These should be soaked for 24 hours in luke-warm water. Sow them 1 cm deep in loose soil and keep in a light situation at around 25°C. A polythene cover will help retain the humidity. Fresh seeds germinate after 4 to 6 weeks. The best results will be obtained in spring, although they may be sown

at any time of the year.

Arrowroot
Maranta arundinacea
Family: Marantaceae

Plants and propagation material are unobtainable. The fine starch of arrowroot is mainly used in baby-food and preparations. It is also suitable for the thickening of sauces and in cooking for invalids. The rhizomes can only be acquired in countries where it is cultivated.

Origin: It is a native of Central America which is still the main area of commercial cultivations. It is also grown to a limited extent in India and Africa.

Habit: The arrowroot grows to a height of 1 to 2 m. The leaves are long stalked, elliptic with a pointed apex and grow either directly from the rhizome or out of the stem nodes. They are arranged in two rows. The flowers are terminal in few-flowered, branched inflorescences. The rhizome itself is covered with scale-like leaves and resembles a huge caterpillar.

Habitat: The plant likes a light sunny position at temperatures from 20 to 30°C and a high air humidity. Since it dies off in late autumn, the rhizome can be kept completely dry in the winter at a temperature not lower than 15°C. Wait until the end of February before planting in fresh soil and then place in a light and warm position. New growth will occur within 4 to 6 weeks.

Soil: A loose, aerated soil containing loam and which drains well is best. A dense retentive growth medium is unsuitable. The fleshy rhizomes will quickly rot in a soil that is constantly wet or permits water accumulation.

Watering: Although arrowroot comes from a warm humid climate one must be careful not to over-water. The plant can absorb a large amount of water and, like the yam, can absorb water when other plants are already beginning to wilt. A slight moisture in the soil is adequate from February till September. From October cease watering altogether, so that the rhizome can mature. When the foliage has gone yellow, the parts of the plant below ground will overwinter in a dry condition. They are not planted in new soil until spring.

Below: the leaves
of the mulberry
tree are the staple
diet for the silk
worm.

Feeding: cf. dasheen (*Colocosia esculenta*) or the sweet potato (*Ipomea batatas*) pp. 54 and 75.

Maturing, harvesting: The onset of growth to maturity takes about 11 months. When the aerial parts go yellow, the rhizomes may be lifted. The tubers, rich in starch, are of commercial importance (see above).

Propagation, cultivation: The arrowroot is propagated by means of cut-back young shoots or parts of the rhizomes. They are set in the plant pot in February and lightly covered with soil. Watered slightly and kept in the light at temperatures around 25°C, they will produce their first shoots after about 3 weeks.

Black mulberry
Morus nigra
Family: Moraceae

Mulberry trees, whether the black or white (*M. alba*) species, are sold by several nurseries and in some garden centres. The fruits of the mulberry are not sold in fruit shops.

Origin: The black mulberry is a native of Persia. Today the tree is grown in those parts of Europe where the climate is suitable for the vine.

Habit: Mulberries form small to medium-sized trees and may grow up to 10 m tall. They are generally monoeceous, with some dioeceous.

The leaves of one tree will often show various forms. They can be ovate, circular, dentate, lobed, heartshaped etc. The flowers are in false spikes. The male ones are cylindrical catkins, the female inflorescences are ovoid and almost sessile. Fruits of the black mulberry are black-red in colour and taste sweet-sourish. They resemble blackberries. The fruits of the white mulberry have an insipid taste. Their leaves are the main food of the silk worm.

Habitat: Mulberry trees should be placed in a sunny spot after the May frosts until the first night frosts in autumn. They will tolerate temperatures close to 0°C when grown in pots, and even lower temperatures when planted out. In winter a cool dry place indoors will suit them; it does not have to be particularly light, since they lose their leaves in autumn.

Dwarf Canary bananas will come into fruit in the greenhouse between 18 months and 3 years according to care.

Soil: Mulberries are happy with any humus soil that drains easily. They are sensitive to cramped root conditions, and therefore need large pots and regular replanting in the first few years.

Watering: The trees need plenty of water as they are fast growing. Maintaining an even moisture in the soil is important. Rain-water should be used. Mulberries are not totally intolerant of lime, but in the long run they will suffer from it. In winter, water only sufficiently to ensure that the roots do not dry out.

Feeding: As they grow rapidly, mulberries need feeding every 14 days from March to August. Cease feeding from September.

Maturing, harvesting: Mulberries will flower throughout the summer, given the right temperatures, if kept under glass. The main flowering period is May–June, the main harvest August–September.

The fruits are generally eaten fresh, tasting pleasantly sweet-sourish.

Propagation, cultivation: Propagation is best effected by taking cuttings in spring. Twigs of pencil thickness are selected and planted deep enough to allow 2 to 3 buds to be under the soil. Keep shady at a temperature of 15°C and roots will form in 4 weeks. Take care not to plant the cutting upside down or it will die. After the first 4 to 6 leaves have formed, transplant.

Dwarf Canary banana
Musa acuminata (syn. *Musa cavendishii*)
Family: Musaceae

Dwarf bananas are obtainable in the fruit trade all year round.
Origin: A native of southern China, they are grown throughout the

tropical and subtropical zones.

Habit: *Musa acuminata* is a perennial herbaceous plant up to 2 m tall, with an irregular rosette of 8 to 12 leaves which can be up to 1.5 m long and 40 cm wide. They are undivided, fresh green, and have a pronounced midrib. The wind can tear them between the lateral veins so that they appear to be pinnate. The stem of the banana is formed by the imbricate leaf-sheath. Through it grows the massive inflorescence which bends downwards once it has reached maturity. The colour of the inflorescence varies from green to deep wine-red. Of the flowers on the pendulous inflorescence only the top rows of pure white ones are capable of forming fruit. All the other flowers are bisexual or male.

The fruits are formed without pollination necessarily taking place. This phenomenon is known as *parthenocarpy*. Each plant flowers only once in its lifetime, bears fruit, then dies. In the meantime, however, it will have produced some suckers which can be removed and used for propagation.

Habitat: The dwarf banana is not so strictly tropical in its temperature requirements as the taller species. Cultivated in a tub, it may be put in a sunny spot sheltered from the wind in the warmest months of the year. It will do best, however, planted in the warm greenhouse. It cannot get too warm for it in the open air, or even under glass. An important factor is to see that the ground temperature does not fall below 12°C in winter. Temperatures down to 10°C will be tolerated, but growth will cease. It is not suitable as a living-room plant as the high air humidity it requires cannot be provided.

Soil: The banana is very fussy about the nutritional content of the soil and its ability to drain easily. The latter should be very humus rich: compost mixed with cattle dung will encourage good growth.

Watering: In view of its rapid growth, the banana needs plenty of water. On dull or cool days, on the other hand, watering must stop immediately, as the plant will not tolerate waterlogging or saturated soil. Water with caution in winter since in cool soil the roots tend to rot. If the water given is too cold it will suffer growth shock and will cease growing for a time.

Feeding: From March until late autumn a weekly organic feed is essential. Dung mixed with water is again the best recipe. Do not feed in winter, since the banana plant grows very little, if at all, during this period and dull weather or too much water will lead to root rot.

Maturing, harvesting: The banana will bear fruit in greenhouse conditions, according to our care of the plant, between 18 months and 3 years. The infrutescence can weigh several kilos. It is left on the plant until the fruits have turned deep yellow. The taste of bananas fully ripened in the greenhouse is delicious.

Propagation, cultivation: Our banana varieties do not produce seeds. They are propagated by means of suckers. These are dug up with their roots and planted in their new position. The best suckers may also, naturally, be left where they are and cultivated in that position. In my experience, suckers with a conical stem have a better yield than those with a cylindrical one.

Packets of banana seed showing a picture of a bunch of bananas on it are sold in the trade. This gives the impression that the amateur can grow his own bananas. This is not possible, as the bananas we know do not form seeds: these seeds are of another banana species which bears fruit which do have seeds, but are unpalatable.

Musa basjoo
Family: Musaceae

This plant is very difficult to obtain. Look out for advertisements in gardening publications.
Origin: Southern Japan.
Habit: This plant is similar to the dwarf banana, but looks in all respects more attractive. Its leaves, which can reach up to 2 m even in our conditions, are easily shredded by the wind and then appear to be pinnate. It will only flower under glass in our climate. The flower

resembles those of other banana species.
Habitat: Its southern Japanese origin tells us that this banana is far less dependent on high temperatures than the other species. If grown in a tub it may be placed in a sunny, sheltered spot after the May frosts and left there until the end of September. It should be protected from the wind, since, having such large leaves, the banana can suffer great damage from it and lose its imposing appearance. In winter it should be put in a light, cool place. 10°C will not harm it.

Soil: cf. Dwarf banana (*M. acuminata*) p. 90.

Watering: As for the dwarf banana p. 90.

Feeding: To avoid excessive growth it is enough to give it an organic feed every 4 weeks. If it gets too large it will need a correspondingly larger indoor overwintering place.

Maturing, harvesting: Neither *Musa basjoo* nor *Ensete ventricolosum* (syn. *Musa ensete*) from Ethiopia, which may be seen planted out in our parks in the summer, bear edible fruit. Both flower only after they are a number of years old. The small bitter bananas, however, contain fertile seeds.

Propagation, cultivation: The propagation of *Musa basjoo* and *Ensete ventricosum* (syn. *Musa ensete*), as far as the seeds are concerned, is the same as for abacá (*Musa textilis*). *Musa basjoo* can also be propagated by means of the suckers it puts out. If they are allowed to remain on the mother plant until they are ½ m tall, they will have roots and may be transplanted without difficulty.

Ensete ventricolosum never produces suckers. Seeds for this plant are always available in garden centres. Sowing in the autumn and winter months is not to be recommended, since these seasons are too dark for the banana plant which thrives on light. All banana seeds should be soaked in warm water at 30°C for about 24 hours to speed up germination.

Abacá
Musa textilis
Family: Musaceae

The useful part of this plant, i.e. the fibres, are only of industrial interest. They are made into paper and coarse textiles.

Origin: South–South East Asia.

Habit: The "textile banana" resembles the fruit banana. It grows to about 3 m. The false stem has a diameter of up to 20 cm. The leaves, which are up to 2 m long, form a dense tuft.

The plant forms small, inedible fruits which contain fertile seeds. The useful parts of the plant are the fibrous bundles which do not rot easily so can be used for textiles which are exposed to moisture.

Habitat: *Musa textilis* is adapted to high temperatures and a great deal of humidity. It can therefore only be grown under glass in our climate, and is not suitable for the living room or garden. It needs a place where it will get plenty of sun. High air humidity is a prerequisite for successful growth. Even in winter temperatures should not fall below 15°C. To ensure it survives the winter, ground heating in whatever form is an advantage.

Soil: *Musa textilis* needs the same

Olive trees are sold in garden centres and nurseries as young plants.

soil as the dwarf banana, cf. p. 89/90.

Watering: The plant must be evenly supplied with water but there should be no accumulation in the soil. The water must also be at the temperature in which the plant is growing. Give less water in winter but a slight even moisture in the soil should be maintained.

Feeding: Feed once a week during the growth period with an organic fertilizer.

Maturing, harvesting: If the plant flowers after a few years the seeds may be used for further stock. It is impossible to make use of the fibres.

Propagation, cultivation: The seeds are set 1 cm deep in humus soil or in pure peat during the spring or summer seasons. The familiar polythene film is placed over the container and it is kept warm (c. 25–30°C) and in semishade. Depending on the age of the seed, germination will follow within 3 to 6 weeks. Those which have not germinated by this time will not do so. The plant also puts out suckers which may be grown in the same way as those of the Dwarf Banana (cf. 89/90).

Olive
Olea europea
Family: Oleaceae

The olive tree is almost always obtainable in garden centres as a young plant.

Origin: The Mediterranean regions are the home of the olive tree. It is, however, grown in other parts of the world with suitable climates.

The fruits of
Opuntia contain a
juicy sweetish
flesh.

Habit: The olive grows as a large shrub or small tree, attaining 12 m in height when old, with a gnarled, partly twisted and often split trunk. The light, many branched crown bears silver-green leaves similar to those of the willow. They are entire, lanceolate and have a short stalk.

The flowers are axillary and resemble those of privet. They appear, depending upon temperature, from February to April; in our conditions as late as July.

The fruits, the well known olives, are formed, in our climate, only on the self-fertilizing varieties. There are oil olives, the flesh of which has a higher oil content, and edible olives, with a greater amount of flesh and less oil. This difference is not visible to the eye, although certain varieties have different leaf shapes. The trees can grow to a great age. Some trees are said to be about 2000 years old.

Habitat: The olive is, without doubt, a 'heliophile'; that is, it needs a great deal of sunlight. It is hardy in the milder regions of the British Isles but otherwise it may be grown in a light living room, since the dry indoor air is like that of its native lands. Even better is a balcony position, or a sheltered place in the garden, where it can remain after the May frosts until the onset of autumn frosts. In winter it needs a light but cool place.

Soil: The soil for *Olea europea* tree should be well aerated and allow for quick drainage of superfluous water. To this end, up to ⅓ sand (building sand) may be included in the potting mixture. The soil may contain lime, for the olive is a calcophile. High peat content is undesirable since it retains too much water. Humic acids also have a negative effect, damaging the roots of the plant.

Watering: The olive tree's great enemy is excessive water. Between waterings the soil ball should almost dry out. It is no use testing by touch to see whether the soil is still damp: the surface can have dried out whereas the interior of the soil ball is still moist. Test by lifting the plant in its container, if grown indoors, and assess by its weight whether it is time to water or not. If in doubt, don't! If too dry, the plant will shed a few leaves, but these will be replaced when it has enough water again. As a result of its origins, the olive tree is very resistant to dryness, whereas wetness will lead to root rot and the tree will die.

Feeding: The olive tree prefers a soil which is not too rich, therefore a feed at the beginning of the growth period and one in summer will suffice. Chemical fertilizers may be used, since a slight mineral accumulation in the soil is tolerated. It is important to cease feeding from August so that the shoots mature.

Maturing, harvesting: Even in our

climate the self fertilizing varieties will produce fruit. These ripen in winter. The green type stay green when ripe. The dark coloured kind obtain this colour when fully ripe.

The olives may be picked and pickled in a 10% brine solution. Change the solution several times until it is clear. The olive stones may be kept and used for sowing in the following spring.

Propagation, cultivation: Cut c. 20 cm long twigs (top end) in April to May and plant to the half way mark in damp sand. Cover with polythene and put in a shady spot. They will root after 2 to 3 months. Transplant carefully to avoid root damage, and place in the sun.

Seeds may be planted in loose sandy soil, about 2 cm deep. They will germinate if fresh after 4 to 6 weeks but it could take from 2 to 3 months. Do not lose patience. The cuttings method is preferable since no grafting is necessary later.

Prickly pear
Opuntia ficus – indica
Family: Cactaceae

The fruits of this cactus are frequently sold in well stocked fruit shops.

Origin: This is unknown. Today *Opuntia ficus-indica* grows throughout the dry zones of temperate regions.

Habit: The bushy or tree-like plant consists of spiny or unarmed, oblong, disk-like or ovate joints. Often a short woody stem develops. The joints can attain a length of 20–30 cm. At the edges of the joints yellow flowers appear, from which hen egg-sized yellow fruits develop. They contain a juicy, slightly sweet, soft fruit flesh. The prickly pear can reach enormous dimensions if well cared for: specimens 1.5 m high and equally wide may be cultivated even in our conditions.

Habitat: In summer it should be placed outdoors or under glass in the dry succulent house. At all events, it should receive as much

Prickly pears can be cultivated under glass or outdoors.

sun as possible. In rainy years it is advisable to provide protection.

It may be kept as a living-room plant, at least while young, since it enjoys dry air. A south facing window would be an ideal position. In winter it needs a cool room, which does not have to be particularly light.

Soil: A non-rich, stony soil, without peat, is necessary. In its natural habitat it colonizes stony, dry areas. If the soil is too rich in nutrients the entire appearance of the plant will suffer and become uncharacteristic.

Watering: Give a thorough watering in spring at the beginning of its growing period. Give no more until the pot or tub has dried out. Water again, and continue thus until into autumn. From winter to spring, unless it is a young plant, it should

not receive one drop of water.

In its homelands the summer temperatures are very high and there are very few rain showers. We cannot offer it warmth in winter, but we can provide dryness. If we water the cactus during winter it will develop an atypical form and there will be no flowering the following year.

Feeding: This presents no problems. An organic feed once or twice in the growth period is completely adequate. Luxurious plants fall more readily victim to disease and their entire appearance changes. Inorganic fertilizer may be given, since the cactus is not sensitive to a slight mineral presence in the soil.

Maturing, harvesting: The fruits on the joints will ripen in the open air during warm summers. They are harvested in autumn. When the

Rice is an annual grass which tillers as early as three weeks after sowing.

berries have turned completely yellow or reddish in colour and the skin yields to the touch, they may be picked.

Wear gloves to do this. At the base of the larger spines are many tiny glochids which, with every movement you make, will penetrate deeper into your hand. If this happens, apply warm candlewax to the spot and when it has cooled it may be possible to remove them along with the hardened wax.

The fruits are peeled with a knife and the flesh eaten with the seeds it contains. In Italy is is believed that the eating of these fruits is beneficial to the kidneys.

Propagation, cultivation: There is no need to raise from seed since there are many specialist firms who supply all manner of cacti including this one. It is offered all year round in all possible sizes.

Rice
Oryza sativa
Family: Gramineae

Rice is easily obtainable but is usually polished, and this type cannot be used for sowing.

Origin: This cannot be established with certainty. India and China have been suggested as possibilities. Cultivation spread from South East Asia to Indonesia, Japan and to the Middle East. It was introduced relatively late to Mediterranean areas, only 300 years ago to North America, and 100 years later to South America. Today it is as a staple food grown in all tropical and subtropical zones of the world.

Habit: Rice is an annual grass which grows vigorously as early as 3 weeks after sowing. Early varieties grow to a height of 50 cm, late maturing varieties to 1.5 m. The culm has leaves which can grow up to a metre long and up to 2 cm wide. The panicle, which bends over when ripe, can be up to ½ m long. The flowering period per panicle is 5–10 days. Self-pollination is common, and in the paddy fields the wind effects this.

Habitat: It cannot ever be too warm for the rice plant in our conditions. It is best suited to cultivation under glass. In the warmest months it can,

Rice is sown from
February to
March and left
uncovered.

however, be placed outside in a sunny, warm place. It cannot be satisfactorily grown in the living room as even in the lightest window it will be too dark for it.

Temperatures of between 25 and 30°C are best for the plant. Since it has a great need for light, it must not be shaded, even when kept under glass. In spring and autumn it is advisable to place the pot on a warm surface. Temperatures below 20°C in these seasons are very damaging to the roots, as inadequate temperatures inhibit their growth. Rice is usually grown in aquatic conditions.

Overwintering is not a consideration since, as an annual grass, rice is harvested in autumn. The seeds are kept dry, cool and dark until the next sowing in spring.

Soil: Rice needs a heavy, loamy soil which is water retentive. Peat must not be included as the plants are very sensitive to humic acids. Loam mixed with up to ⅓ lime free sand is the correct mixture. Organic components in the soil would rot, since the rice stands in water for the best part of the year.

Watering: During the entire growth period use only rain-water or softened water. The use of tapwater will introduce too many mineral salts into the soil, particularly when the water consumption is at its peak in warm temperatures. When the culms go yellow, the watering can be cut down or stopped altogether.

Feeding: As soon as the young plants are about 15 cm high, start to feed. In my experience a fortnightly feed of organic fertilizer e.g. manure dissolved in water, is beneficial. Cease feeding when the plant has matured, i.e. when the culms start to turn yellow. In normal summers this will be in the middle or at the end of August.

Maturing, harvesting: At the end of the growth period the plant begins to flower. To get a good crop, it is advisable to grasp the lower part of a number of inflorescences, and shake them together to spread the pollen around. Rice is normally wind pollinated, but this is only successful in dry weather. The panicles are at first green and become yellowish as they ripen. They are harvested when the single grains feel hard when touched. Store them cool, dark and dry during winter.

Propagation, cultivation: Rice is only propagated by seed, so one must have fertile material. Most rice we buy for culinary purposes is useless, as when it is prepared the embryo is removed. Health food shops sell unpolished rice, and if it is not too old, a proportion of it will germinate.

In February/March the seeds are sown thickly without covering on a humus bed. This bed is not inconsistent with the above recommended soil mixture. In the germinating stage waterlogged conditions are not provided, and the roots of the young plants can more easily make their way through the soil. Placed in a light position at 20 to 25°C, the seeds will germinate after 8 to 14 days.

As soon as the young plants are c. 10 cm high, they may be transplanted. Fill a container, pot or tub without a drainage hole, half full with the loam mixture. Top up with water to several centimetres above the loam level. Carefully free the roots of the young plants from the soil and plant up to the root collar in the new growing medium.

The water level, a few centimetres above soil level, can be maintained until the plant turns yellow. It will not hurt to let the water level sink from time to time, as this will re-aerate the soil. The water temperature should not drop much below 20°C during the growing period.

Passion fruit or purple granadilla
Passiflora edulis
Family: Passifloraceae

The fruits are sold fresh in early and high summer.

Origin: A native of Brazil, *Passiflora edulis* is today grown in Africa, Australia, South Africa, New Zealand and Hawaii.

Habit: It is a climbing liana, several metres long. The leaves are dark green and three lobed. The large white flowers arising from the leaf axils are self-pollinating. The fruit grow very quickly to hen egg size, but hang on the plant for several months before they turn purple-violet.

The slightly slippery flesh is sourish-sweet in taste. The many

A delicious juice is extracted from the fruits of *Passiflora edulis.*

seeds are eaten with it. When pollinating the flowers, one must wait until the anthers open, and the pollen, which has a waxy consistency when the flowers open, has become powdery.

Habitat: The plant tolerates direct sunlight whether under glass or in a tub outside. It may be put outside after the May frosts until towards the end of September in a place sheltered from the wind. It will develop best planted in a greenhouse. It is not really suitable for a living room if grown as a fruit bearer, as it becomes too extensive. It may, however, be grown there as an ornamental flowering plant. In winter it needs a light place which should be on the cool side, c. 10 to 15°C.

Soil: Standard potting material mixed with ⅓ peat has proved adequate.

Watering: On account of its rapid growth and dense foliage *Passifolia edulis* needs plenty of water. No dry periods must be permitted when the fruit has set, or it will be shed. In winter an even but slight soil moisture must be maintained, depending on the temperature of the room.

Feeding: The passion fruit plant responds well to organic feeding. This should take place at 4 weekly intervals until the beginning of the flowering period. More frequent feeding at this stage would produce too much foliage and fewer flowers. When the fruits have started to form the feeds may be increased to fortnightly. Cease feeding at the end of August so that, with failing light, there will be no new growth.

Maturing, harvesting: The fruits are ripe when the skin is dark purple. The skin is not edible, but

The flower and fruit of the giant granadilla can be 25 cm long and up to 20 cm thick.

the seeds are eaten with the fruit flesh (strictly speaking these are arils). They can also be used in jams or fruit salads.

Propagation, cultivation: Remove the seeds from a fruit you have purchased and clean off any of the mucilaginous flesh. Sow in spring in a humus soil or peaty bed, but do not cover with soil. Cover with polythene. Keep in semi-shade at c. 20 to 25°C. Germination should begin after about 14 days to 3 weeks.

Passiflora edulis may also be propagated by taking cuttings. A not fully matured shoot about 20 cm long is cut below a leaf axil and set in damp peat. The use of a rooting powder will encourage development and reduce the danger of decay. Kept at 20°C in a shady but light position, the cutting will start to grow roots in about 3 weeks.

Giant granadilla
Passiflora quadrangularis
Family: Passifloraceae

The fruits are only occasionally to be seen on the market.

Origin: It is a native of tropical America.

Habit: The giant passion fruit is also a climber. It has acute, ovate, light-green leaves the size of the human hand. The branches can be several metres in length. They keep dividing so that one plant can be several metres high and wide. The tea-plate sized flowers produce fruit which may be up to 25 cm long and 20 cm thick.

Habitat: *Passiflora quadrangularis* will only grow under glass under subtropical or tropical conditions. Cultivation in a tub is possible, but it would have to be a large one. It is not possible to grow it in the living room since the plant enjoys direct sunlight. The temperatures can exceed 30°C. In the winter it is kept somewhat drier at temperatures not below 15°C.

Soil: cf. the passion fruit (*Passiflora edulis*) p. 99/100.

Watering: cf. the passion fruit (*Passiflora edulis*) p. 99/100.

Maturing, harvesting: The giant passion fruit will not flower in our

conditions unless grown in a large greenhouse. The inside of the fruit is not eaten, but the thick fleshy skin is consumed as a compôte or vegetable.

Propagation, cultivation: *Passiflora quadrangularis* can only be grown from seed. For the procedure cf. *Passiflora edulis* p. 99/100.

Avocado
Persea americana
Family: Lauraceae

Avocados are available all year round.

Origin: It is a native of Central America but is cultivated worldwide in suitable climates.

Habit: In its homelands, the avoca-do is a tree up to 20 m in height, with pointed elliptical leaves, dark green in colour. It has terminal racemes with very small green flowers. From these develop the well known pear-shaped (in some varieties globular) green to black-ish-seeded berries. There are three types: the Mexican (var. *drymifolia*) which are small; the West Indian, which are large with smooth skins, and the Guatemalan, also large with warty skins. With its high fat content the fruit is to be considered more as a vegetable than a fruit.

The Mexican type is said to be the least sensitive to cold.

Habitat: The avocado may be grown in a tub or in the green-house. It should not grow outside except in the warmest months, in a warm place sheltered from the wind. Under glass it will tolerate direct sunlight even in the summer months. The temperature may exceed 30°C.

In winter it needs a light but not too cool position. Temperatures should not fall below 15°C. If during the year the tree is beginning to take up too much space it may be cut back. It will put out new shoots from the leaf axils and can also be grown as a shrub.

Soil: The avocado can grow in standard potting material made less rich by an addition of sharp sand.

Feeding: A feed every 4 weeks during the growth period is enough for adequate growth. Since the avocado rarely flowers as a potted plant, rapid growth is not really desirable. This plant is most attractive for its foliage, and makes an unusual pot plant.

Watering: It is sensitive to waterlogging and mineral accumulations in the soil, so allow it to dry out somewhat between waterings and use only organic fertilizers. The latter may be purchased at any garden centre. Cease feeding in September, otherwise the tree will go on growing and its shoots become so thin on account of poor light that they cannot support themselves. They would have to be cut off in spring anyway.

In winter ensure only a slight moisture in the soil. At the same time, even in the cold season it must be watered every few weeks, since it keeps its leaves.

Maturing, harvesting: The amateur grower is not likely to be able to obtain an avocado crop, since several trees of different groups are necessary for this purpose. All avocados are protogynous, i.e. the stigmas mature before the anthers. Self-fertilization is therefore not possible. When the stigmas of one group are ready for fertilization, the pollen of the other group must be ripe.

Propagation, cultivation: The large stone is removed from the fruit and set pointed end upwards 2 to 3 cm deep in peaty soil. Most of the stone is left exposed. Water lightly and enclose the pot in a transparent polythene bag. The pot should be kept light but not in direct sunlight at c. 20 to 25°C.

The stone must be partly exposed so that it can split when germination sets in. Planting thus will also ensure that the young shoot does not grow twisted and eventually die. When the fourth leaf appears, remove the plastic bag and cultivate normally. Do not remove the stone by force but wait until it rots away.

The date palm is
grown for
ornamental
purposes, it will
not bear fruit in
our climate.

Date
Phoenix dactylifera
Family: Palmae

The fruits of the date palm are obtainable all year round. When in season, fresh fruits are sometimes sold in supermarkets and fruit shops.

Origin: Wild forms of the date are unknown. It is an ancient cultivated plant of great commercial importance.

Habit: The date palm achieves a height of 15 to 25 m or more and has a large trunk covered with the remains of leaf stalks. The 40 to 50 pinnate leaves are up to 4 m long and form a dense crown. They remain on the plant all year round.

Phoenix dactylifera is dioecious. The male and female flowers are arranged in separate much-branched panicles which grow from the leaf axils.

Early in their development these panicles are enclosed by a fibrous spathe. The female flowers are loosely arranged, whereas the male ones are grouped close together. Normally date palms are wind pollinated, but in cultivated plants the female flowers are artificially pollinated; the male inflorescences are cut out and placed above the female panicles. Since the pollen is long-lasting it can be obtained commercially. From the female flowers develop the longish single-seeded fruits, the well known dates. These are green at first, then yellow, red, or even blackish. The fruit flesh in some varieties is soft, high in sugar content and scenty; in others harder and drier.

Habitat: The date palm likes much warmth and light. It can never be too warm for it in our climate. In summer it may be grown in the living room or garden. Dry air will not harm it, indeed this is preferable. It should stand in direct sunlight if outside in summer; from autumn to spring it can stand in a light window. Temperatures may be allowed to fall to 10°C.

Soil: The plant is not too difficult in this respect. Humus soil with a ⅓ loam admixture is suitable. The fleshy roots are sensitive to too much water, so the soil must drain easily. When replanting, make sure that the new soil is well pressed down, using a piece of wood to ensure this, if necessary. Do not replant until the roots are lifting the soil ball out of the pot.

Watering: The soil ball should be slightly moist all year round. In winter it should be kept evenly moist, the degree depending on the temperature of the room. If replanted annually in fresh soil, the palm may be given tap water, but make sure that it is always at room temperature.

Feeding: It is only necessary to feed every 2 months, since we do not want the plant to grow too quickly. If this happens it will not fit into any room after a few years.

Chemical fertilizers may be used as the date palm will tolerate a soil mineral content of up to 3%.

Maturing, harvesting: The tree will not bear fruit in our climate. The most northerly site where it will fruit is Southern Spain, but the yield is low. There they are grown for their long, pinnate leaves.

Propagation, cultivation: Date palms are propagated over here by means of seeds. The seeds of dry dates germinate more easily than those of the fresh fruit. The latter are probably not quite ripe when harvested. Soak the stones for 2 to 3 days in lukewarm water, then plant flat 1 to 2 cm deep in the soil. Keep in a warm light position and water slightly. According to the age of the seeds, germination will take place after 2 to 6 months. Date Palm seeds can still germinate after 2 years.

At first the young plants have pointed lanceolate leaves, then comes an intermediate form which have forked apices. The next leaves will be the characteristic pinnate kind. When transplanting the young plant, make sure that the stone still attached to it is not lost, as it still provides the plant with nourishment at this stage.

Planted in a sunny, warm position, preferably against a south-facing wall, the cape gooseberry will produce fruits.

are surrounded by an inflated dry calyx. In this respect they resemble the Chinese lantern plant *Physalis alkekengi* which is grown as an ornamental plant in our gardens. The cape gooseberry attains a height of 100–150 cm.

Habitat: If kept in a sunny, warm place on a south-facing wall, the plant will fruit reliably in our climate. Wind is not good for it, since its herbaceous stems easily break. If you wish to keep the plant for several years, it needs a light but cool place in winter. It must not be allowed to grow in winter, as our poor light conditions would result in long shoots which are not self supporting and which would have to be taken off in spring.

Soil: Any loose humus soil is suitable.

Watering: A great deal of water is lost through the soft green leaves in summer and this must be continually replaced. *Physalis* will not tolerate drying out of the soil ball. Since the leaves are not shed in winter, it also needs a constant slight soil moisture in this season. It is best to water just when the leaves begin to wilt.

Feeding: To obtain good growth and a satisfactory crop, a weekly organic feed in the vegetation period is necessary. Cease feeding from the end of August. If the plant gets too big for its overwintering position it can be cut back to half its size. The branches from which fruit has been picked can certainly be

Cape gooseberry
Physalis peruviana
Family: Solanaceae

These can be bought in more exclusive shops during spring and summer.

Origin: It is a native of South America. There are now large cultivations in South Africa.

Habit: The cape gooseberry is not related to the European gooseberry *Ribes uva-crispa*, nor does it resemble it in any way. It is an annual or perennial herb with soft, heavy, ovate, pointed, medium-sized leaves. In the leaf axils, particularly near the growing point the pale yellow flowers appear which may be fertilized by their own pollen.

Cherry-sized berries develop. They are yellowish when ripe and have a sweetish-sour taste. They

removed, since the plant will only bear fruit on new shoots.

Maturing, harvesting: The fruits are ripe when the calyx which surrounds them is dry and papery. The fruits will then be pale yellow. They may be eaten raw or, if the quantity merits it, made into an excellent jam. The tiny seeds inside the fruit are also eaten.

Propagation, cultivation: The seeds are obtainable in gardening shops. Sow finely on a bed of standard potting material but do not bury. Cover with a polythene bag to retain air humidity. Place in a shaded position at 20 to 25°C.

Plant out singly after the appearance of the second leaf pair. You will have more plants than you need. Sowing in spring brings the greatest success. If well cared for, cape gooseberries planted in February will flower in the summer.

Pepper
Piper nigrum
Family: Piperaceae

The dried fruits of this tree are sold whole or ground into powder. Black and white pepepr come from the same plant and are merely harvested at different times and prepared differently.

Origin: Probably the southern part of the Indian subcontinent. Today pepper is cultivated everywhere in the tropics where there is sufficient rain.

Habit: The pepper is a climbing plant, woody in its lower, but green in its upper parts. The pencil-thick knotty stem is pliable and forms adventitious roots with which it can, like ivy, climb to 10 or 15 m. The leaves are dark green with pronounced venation, are thick and opposite with a short stalk. The flowers are very small and form catkin-like spikes. They are usually unisexual, and cultivated plants are mostly monoecious. One spike will produce 20–30 peppercorns. At first they are green but turn red as they ripen.

Habitat: Pepper requires a light position all year round, at temperatures not below 18°C. It is not possible to grow it in the living room. It may only be put outside in

the very warmest weeks of the year, and then only in a sunny, draught free spot. It can already be too cold for it by mid-August. In winter it needs a light position and temperatures no lower than 18°C.

Soil: The plant needs a drainable, loose soil with high humus content. Potting material with too high a peat content which retains water for a long time is not suitable. Standard potting material with ⅓ sharp sand is a good mixture.

Watering: The pepper plant requires a slight, even moisture in the soil. Lengthy wet periods or waterlogging will lead to root rot in a very short time. Greatest care must be taken in winter. A watering followed by a long sunless period spells disaster. The only remedy is to stand the pot on top of a radiator so that the temperature around the roots is increased. Long term, the pepper will only tolerate rain-water since it likes the soil to be slightly acidic.

Feeding: From March into September the *Piper nigrum* can be fed every 4 weeks. In my experience organic fertilizer (dung dissolved in water) is more successful than the chemical variety.

Maturing, harvesting: About half a year separates flowering and harvesting. If the fruits are harvested before fully ripe, when they are just turning red, black pepper is the result. The peppercorns are picked and dried. White pepper comes from fully ripe fruits. In this case they are soaked in water until the fruit flesh gets detached. This reveals the greyish-white seeds which are then dried. White pepper is hotter than black.

Propagation, cultivation: Fertile seeds must be obtained from their countries of origin. They must be sown immediately, whatever the time of the year, as they quickly lose their power to germinate. Covered with soil and kept light and warm they germinate after 2 to 3 weeks.

The pepper plant is, however, mainly propagated by making cuttings. Strong twigs or root suckers are used for this purpose. These should be c. 30 to 40 cm long; the top half of any leaves should be cut off, and the twigs then covered with soil. Cover with a polythene bag to retain the air humidity and keep evaporation to a minimum; this is necessary because there are as yet no roots to supply water. Leave in a light, warm place but protected from the sun for 4 to 6 weeks. As soon as the new growth is visible, the polythene bag may be removed.

Pistachio
Pistacia vera
Family: Anacardiaceae

Roast pistachio nuts are sold all year round in food shops.

Origin: A native of Central Asia and the Near East.

Habit: In its homelands the pistachio tree grows up to 10 m. It has greyish green pinnate leaves. Inconspicuous inflorescences grow from the leaf axils. The tree is dioecious. To obtain fruits one must therefore keep both male and female trees. Pistachio nuts are 2–3 cm long and egg shaped, thin-shelled and possess green cotyledons.

Habitat: Since the pistachio is also grown in the Mediterranean region, it should be grown either in the succulent house or, in summer, in the open air in a warm, wind

sheltered spot. It tolerates direct sunlight. The upper temperatures have no limit; in winter 10°C is adequate, and its position does not even have to be particularly light, since the tree sheds its leaves in late autumn. It is possible to grow it in the living room as it is adapted to dry air. In wet summers it must be protected from constant dampness or root rot will set in.

Soil: A not too rich, well drained soil is best for this plant. The draining quality of the soil is important to avoid root rot.

Watering: Growing in dry conditions, the pistachio should only be given water when the soil ball has dried out. If you omit to water at the right time, it will drop some of its leaves, but these will be replaced once it has received water again. In autumn all the leaves are shed, so that in winter only a slight soil moisture is necessary – just enough to ensure that the roots do not dry out.

Feeding: One feed at the beginning and one in the middle of the growth period is enough for the pistachio tree. Its relatively large root system exploits the nutritional content of the soil to the full, so that the necessary transplanting is enough to ensure that the plant is adequately supplied.

Propagation, cultivation: *Pistacia* is propagated by means of seed. Unfortunately, the pistachio nuts on sale here have been roasted and will not germinate. Seeds could be

The strawberry guava is a short-stemmed shrub.

obtained from Mediterranean regions.

Having obtained some, soak them for a few days in lukewarm water, then place them on the surface of a pot full of slightly moist cactus potting mixture and press lightly down. They should be only half in the soil. Germination will take about 4 weeks at c. 20°C. Do not cover with polythene as this will create over-warm conditions.

Strawberry guava
Psidium cattleyanum
Family: Myrtaceae

Fruit or plants are hard to obtain. Keep your eyes open for advertisements in horticultural publications. Fruits may sometimes be found.

Origin: Brazil. This plant is much hardier than the real guava, but has much smaller fruits. It is cultivated mainly locally, but is also to be found in southern California and on the Gulf Coast (often as an ornamental shrub).

Habit: The strawberry guava is a shrub-like, short-stemmed plant with opposite, almost ovate, shiny green leaves. The flowers are formed in the leaf axils, as in the guava (*Psidium guajava*). They are equally attractive with their long protruding stamens. From them develop cherry-sized, red, sourish berries which may be eaten raw.

Habitat: *Psidium cattleyanum* is hardier than its relatives. It may be kept from spring to late autumn in a sunny sheltered spot outside. It will not survive frost. In winter it needs a light but cool place. Temperatures below 10°C will only be tolerated if the soil ball is only slightly moist.

Soil: cf. Guava (*Psidium guajava*) p. 111.

Watering: The strawberry guava appreciates regular watering, but will not suffer if the soil ball dries out for a brief spell. Regular watering is important, however, when it is forming its fruit, for if deprived, it will drop them. The cooler the temperature in its overwintering position, the drier the soil must be.

Feeding: cf. Guava (*Psidium guajava*) p. 111.

Maturing, harvesting: When the berries are dark red and start to fall, they are ripe and may be eaten raw.

Propagation, cultivation: cf. Guava (*Psidium guajava*) p. 112.

The fruit flesh of the guava possesses a wonderful scent, but its taste is somewhat disappointing.

Guava
Psidium guajava
Family: Myrtaceae

Fruits are occasionally obtainable in shops in spring and summer.

Origin: It is a native of tropical America, but is now grown in warm countries throughout the world.

Habit: In the above mentioned countries it will grow to be a 10 m high shrub or tree. It has opposite, entire leaves the size of those of the cherry tree. The lateral nerves are distinctly impressed. The fairly large white flowers are arranged singly in the leaf axils. They have numerous long protruding stamens. Yellow pear-shaped berries with a waxy skin develop from the self-fertilized ovaries.

Habitat: The guava likes to be in the sun. Temperatures may rise to 30°C and more in summer when it may be kept in the open air in a sheltered spot. It tolerates a wide range of temperatures. In winter 10°C is warm enough, but the plant will not grow at that temperature.

It may be grown in the living room or conservatory, but in winter should be placed in a cooler room. It cannot go without the maximum possible light. We must never forget that in the tropics and sub-tropics there is at all times more light than at our latitudes.

Soil: The guava is not too difficult in the matter of soil. Any soil with a reasonable humus content will do. The only proviso is that it should not turn into mud when watered, otherwise it will cut off the roots' oxygen supply.

Watering: The guava's ability to resist drought is impressive, but an even supply of water will ensure undisturbed growth. Since the tree retains its leaves during winter, it must also be watered in this season, but less as the temperatures fall. It is sensitive to wet conditions in winter.

Feeding: The tree bears fruit, if kept in the greenhouse or planted in the garden, from the 3rd or 4th year.

Feeding every 4 weeks will ensure a good crop of fruit. Young

The pomegranate
displays beautiful
axillary flowers
(right). The

mature fruit (far
right) is squeezed
in the hand and
the juice drunk.

plants, which are not yet capable of flowering, are given an organic feed every 6 to 8 weeks. Do not feed outside the growth period.

Maturing, harvesting: The fruits which follow from flowering grow quickly to the size of a small pear. They then hang for some weeks on the shrub or tree before they turn light yellow. They are ripe when they yield to pressure from the finger, or when the tree itself begins to drop them. Under the thin skin there is pink, wonderfully fragrant fruit flesh containing many seeds. In flavour it is something between a quince and a raspberry. The fruits will not keep and must be used immediately. An excellent jam may be made with them.

Propagation, cultivation: The seeds must be placed on the surface of a pot full of the soil described above. Kept in a shady place they will germinate in 3 to 4 weeks. Covering with polythene will retain the air humidity and encourage germination.

If you have obtained a cutting from a botanical garden, remove the leaves, without harming the buds present in the leaf axils. Keep it shady at c. 20 to 25°C. The air-layering method should be employed here too (see p. 119). If a rooting powder has been applied, the chances of it taking are increased. After 4 to 6 weeks the first leaf tips of the new growth should be visible.

Pomegranate
Punica granatum
Family: Punicaceae

Fresh pomegranates are sold in fruit shops from spring to autumn.
Origin: It is distributed from Asia to the Mediterranean countries and today is cultivated in any region with a suitable climate.
Habit: The pomegranate grows as a tree or shrub. It has small, ovate, lanceolate fresh green leaves. The new growth is red. In our conditions the tree sheds its leaves in autumn, but in its homelands they are often retained.

The beautiful red flowers are formed in the leaf axils towards the apex of short branches. They are campanulate, with a persistent calyx. The petals are folded in bud. From these develop the apple sized fruits known as pomegranates. It is often said that the fruits of this tree gave their name to the Spanish town of Granada, and there is, indeed, a pomegranate in its coat of arms. The generic name stems from the Latin *malum punicum* (apple of Carthage).

Apart from the fruit bearing varieties, there is also a dwarf form, a small plant which flowers reliably and occasionally forms fruit.
Habitat: The pomegranate should be placed in a sunny spot, sheltered from the wind, from spring until late autumn. It is also suitable for keeping in a flat, if it can be moved on to a balcony in summer. It will

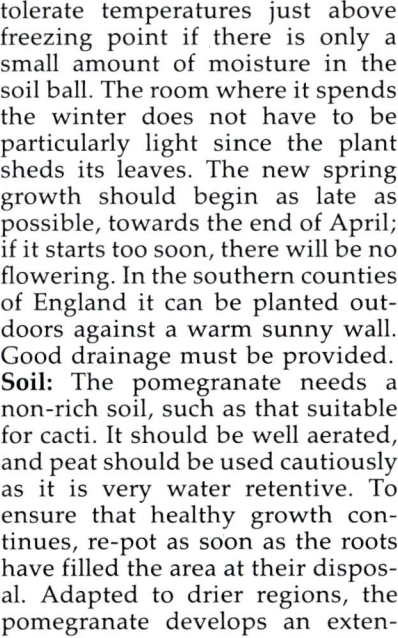

tolerate temperatures just above freezing point if there is only a small amount of moisture in the soil ball. The room where it spends the winter does not have to be particularly light since the plant sheds its leaves. The new spring growth should begin as late as possible, towards the end of April; if it starts too soon, there will be no flowering. In the southern counties of England it can be planted outdoors against a warm sunny wall. Good drainage must be provided.

Soil: The pomegranate needs a non-rich soil, such as that suitable for cacti. It should be well aerated, and peat should be used cautiously as it is very water retentive. To ensure that healthy growth continues, re-pot as soon as the roots have filled the area at their disposal. Adapted to drier regions, the pomegranate develops an exten-

sive root system, so the container must be correspondingly large.

Watering: In the growth period the pomegranate needs plenty of water. Between waterings, the soil ball should dry out slightly, so that there is no water build-up. In winter keep the soil ball slightly moist. The drier it is, the lower the temperatures may be in the room where it overwinters.

Feeding: Feed organically every 4 weeks from the onset of growth to the end of August. This is sufficient; plants that are too extensive will flower badly.

Maturing, harvesting: Only older specimens of the pomegranate tree will bear fruit in warm summers over here. More often than not, the quince-like fruits do not manage to ripen before winter sets in, unless grown under glass. Ripe fruits may be squeezed in the hand and then

In our climate sugar cane must be cultivated under glass. Unfortunately it will not produce sugar.

the juice drunk.

Propagation, cultivation: This is usually be means of cuttings. In spring, take cuttings about 20 cm long and plant in loose soil. They will root, kept at 20 to 25°C, in 4 to 6 weeks.

Sugar cane
Saccharum officinarum
Family: Gramineae

Sugar cane cannot usually be bought as a plant in the trade. Occasionally it is possible to obtain a cutting from a botanical garden. **Origin:** Sugar cane was first cultivated in India and spread eastwards from there. It was brought to the west by the Arabs, then by the Portuguese and Spanish to the West Indies and Central and South America.

Habit: Sugar cane is a tall grass, several metres high, with a massive, many-noded culm, the diameter of which can be several centimetres. It is light green, but reddish and red varieties occur. The plant is perennial and branches grow from the lower half. The internodes are covered by leaf sheaths over most of the length of the culm. The leaves are from 30 cm to 2 m long and 3 to 6 cm wide according to variety and have razor-sharp edges, so that careless handling can result in cuts. In our climate sugar cane will not flower. The flowers are arranged in dense panicles.

Habitat: In our climate sugar cane must be kept under glass all year round. It is not possible to grow it in the living room. It could thrive in summer against a particularly warm house wall. It does best in tropical conditions at temperatures between 23 and 30°C. They should not fall below 20°C in winter, even when the soil ball is relatively dry.

The plant requires as much light as possible all year round, so no shading is required, even under glass, in summer. Temperatures below 20°C will not kill the plant but growth will be considerably checked. Night temperatures may fall a few degrees below those of the day.

Soil: Sugar cane is not particularly demanding in this respect. It should be humus soil and well drained. Standard potting material is fine.

Watering: *Saccharum* needs plenty of water as long as the temperatures are high. In tub cultivation temporary water logging will be tolerated. In cold months, however, reduce the water so that there is only a slight moisture in the soil. In view of the high water consumption,

114

The fruits of the lulo make a delicious jam.

rain-water should be used to avoid mineral accumulation in the soil.

Feeding: Since it grows quickly, the plant needs plenty of feeding. For this reason feed from the beginning of the growth period to the middle or end of September at intervals of 10 to 12 days. A mixture of manure and chemical fertilizer has proved successful. Put a handful of cattle dung and half the quantity of chemical fertilizer into a watering can, let it stand for a few days, stir and use. This mixture has proved its worth for years with all the plants mentioned in this book.

Maturing, harvesting: Information would be superfluous, as the plant does not produce sugar in our climate.

Propagation, cultivation: Sugar cane is generally propagated by means of cuttings which may be taken from any part of the culm. Each should have a node. Plant flat or sloping in standard potting material so that the node is just below the surface. Covered with polythene and kept light but not in direct sunlight at c. 25°C they will produce roots in just a few days. They may be transplanted after about 4 weeks.

Lulo

Solanum quitoense

Family: Solanaceae

Fruits and seeds are not on sale over here. Material can only be obtained from the countries where it grows or from a botanical garden.

Origin: It is a native of the South American Andes. Its cultivation has spread via Columbia and Ecuador to the east coast of South America wherever the climate permits.

Habit: A coarse perennial or sub-shrub which can be up to 2 m high and wide. The leaves are fresh-green, broadly ovate, up to 25 cm long and broad, coarsely crenate, armed with stiff bristles on both surfaces and often also (like the leaf stalks and stems) with 2 cm long spines. The veins are often purplish.

Flowers are small, similar to those of the tomato and arranged in axillary cymes; they are self-pollinated and the tomato-like fruits develop from the ovaries. They are the size of cherries, at first green then orange or red at maturity; the fruit stalks are also spiny.

Habitat: The lulo needs to be kept under glass all year round in our climate. Only in the warmest time of the year could we possibly put it outside, against a sunny wall. It is possible to grow it on a light window sill indoors, but it requires sensitivity to the plant's needs. The many thorns that lulo possess,

Flowers and fruits
of cocoa develop
directly on older
branches and on

the trunk. The
leaves seen here
belong to
Piper betle.

which prick us and hang on to us when we go anywhere near it, deter predators. In autumn the upper parts become dry. This condition sets in very quickly, but this does not mean that one has made a mistake during cultivation.

Soil: The plant is not particularly fussy. Standard potting mixture is perfectly all right and will provide for its needs in the growth period. Long periods of soil wetness or waterlogging do not agree with the plant.

Watering: *Solanum quitoense* needs a constant even supply of water. If the leaves wilt too much, they will not recover. The plant will not tolerate cold water, so allow tap water to stand until it is at room temperature.

Feeding: If the plant has plenty of space to grow, feed a dung solution every 4 weeks. It will grow then into a very interesting shrub.

Maturing, harvesting: When the berries have gone dark red they are ripe. They can remain for weeks on the branches which will now gradually become drier in appearance. Keep berries intended for propagation in a dark, not too cold room over winter. If there are enough fruits, an excellent jam may be made.

Propagation, cultivation: If you have fertile seeds, treat them like tomato seeds. Cover slightly with soil, and keep light, warm and slightly moist. Germination starts in 2 to 3 weeks. In order to get the benefit of a full growth period, sow as early as February.

Cocoa
Theobroma cacao
Family: Sterculiaceae

Seeds and plants are not easily obtainable. A journey to the Caribbean is necessary to acquire seeds. Some cocoa plants in botanical collections produce seeds and if one is lucky, it might be possible to obtain some.

Origin: This is today considered to have been in South America, in the Amazon region. Cocoa was introduced into Europe by the Spaniards. Apart from in South America, it is now cultivated in Sri Lanka, Indonesia, New Guinea and Central and West Africa.

Habit: Mature trees are 4 to 8 m tall and their trunk can be up to 25 cm in diameter. The leaves are up to 30 cm long, pendulous, initially red, later dark green and parchment-like. The flowers are directly borne on stronger branches and mainly on the trunk. They are about 10 mm in diameter, yellowish, long-stalked and have a purple calyx. In its natural habitat the tree flowers all year round. The fruits are egg to cucumber shaped, ribbed, up to 30 cm long, yellowish, purple or brown. The ellipsoid seeds are up to 25 mm long and

Cocoa seeds are
the raw material
for cocoa and
chocolate.

embedded in a mucilaginous pulp; they yield cocoa after fermenting and roasting.

Habitat: *Theobroma cacao* can only thrive well in tropical regions and therefore must be kept in the hot house. It requires temperatures up to 30°C in summer and they should not drop below 20°C in winter. The 20°C mark is necessary in the soil ball, therefore I place my cocoa plant directly over the radiator in the cold season.

As an undergrowth plant of the tropical rain forest, the plant needs slight shading from the sun in summer, but not in the other seasons. An attempt to grow it in the living room is doomed to failure since it requires high air humidity all year round.

Soil: It does best in a slightly acid soil which should contain humus and peat. A good idea is to put 2 cm of sharp sand at the bottom of the pot to assist drainage, as the cocoa roots react badly to too much water lingering in the soil. It is nevertheless essential to see that the potting mixture retains enough moisture; alternating wet and dry conditions will kill the plant.

Watering: The cocoa plant requires lime-free water at room temperature. It is very sensitive to too much water, especially at temperatures around 20°C. Even soil moisture is essential for healthy growth. Reduce watering in the cold season. Cocoa is less harmed by short dry periods than by overwatering. Frequent spraying should be carried out.

Feeding: Cocoa will respond well to feeding once in four weeks during the period from March to September. The early start to feeding is explained by the plant's position in the hot house. There the increasing amount of light will trigger off early vegetative growth which will continue almost until winter. The fertilizer mixture described on p. 53 may also be used for this plant.

Maturing, harvesting: Information here would be irrelevant, since cocoa will only fruit in the large glasshouses in botanical gardens where it is planted in the ground or at least in extremely large tubs.

Propagation, cultivation: Cocoa beans lose their ability to germinate very quickly and must therefore be sown as soon as they are obtained, regardless of season. Plant in the soil mixture described above, having soaked them for about 25 hours in lukewarm water. Plant as deeply as the beans are broad.

Fresh seed will germinate after 14 days. They must not be exposed to the sun at this stage. A light position with 80% air humidity and temperatures around 25°C are the ideal conditions. Seeds which have not germinated after 14 days are dead. Plant singly after the appearance of the second pair of leaves. Take care that a small soil ball remains on each as you do so. The cocoa root must not be bent during this procedure.

Vegetative propagation methods available to the amateur are air-layering or the taking of cuttings. The first method involves notching the shoot with a knife 5–10 cm below its tip and tying sphagnum moss around this exposed spot. Tie polythene round to keep the moss moist. After a few weeks roots will develop and the stem can be severed below these and planted.

Cuttings are taken from mature shoots of lateral branches. They should be c. 20 cm long. Plant in sharp sand, as only in this medium will they form normal roots. Cover in polythene and place in a shady place at 30–35°C and roots will form after a few weeks. Other methods of propagation are of no interest to the amateur. Both methods described must be applied during the growing season.

Vanilla
Vanilla planifolia
Family: Orchidaceae

Since the vanilla plant is propagated exclusively by means of cuttings and is not to be found in the average garden centre it is difficult to obtain. As it is a member of the orchid family, it may be obtained from a specialist orchid grower.

Origin: It is a native of the tropical rainforests of America. Last century young plants were introduced into Java via the botanical gardens of Paris and Antwerp. Since the insect necessary for the pollination of this plant, a moth, did not live in its new habitat, early attempts to grow it failed. With the development of a technique for manual pollination, its cultivation spread, mainly in Indonesia and Madagascar. The genus Vanilla comprises over 100 species. Out of more than 10 000 orchid species *V. planifolia* is the only one of commercial use (other than as an ornamental).

Habit: The vine-like plant posses-

ses a fleshy stem as thick as a pencil; it climbs up to 10 m or more, each node bears a leaf and opposite, a root. The numerous sessile fleshy leaves are 12–20 cm long and up to 5 cm broad. The inflorescences are axillary, clustered and several-flowered. The green to yellowish flowers, which are produced in succession, are of a complicated structure and not self-pollinating. In cultivation, the powdery or granular pollen has to be transferred from the incumbent anther artificially. The fruit is a bean-shaped fleshy pod up to 15 cm long. It will develop over a period of 4–6 weeks but it takes a further 6 to 9 months to reach maturity.

Habitat: Coming from humid tropical regions, the *Vanilla* must be grown in the hot house. Shaded slightly, it will not suffer from the highest summer temperatures. It also has an amazing ability to adapt to winter temperatures and will survive perfectly at around 15°C. A precondition, however, is an almost dry soil ball.

It is not possible to grow it in the living room, nor can it be placed outside. Air humidity above 70% is absolutely essential. It must also be given some means of support, as it is a climbing orchid with very long stems.

Soil: The growth medium must be purchased from a specialist firm. It is a coarse material, very well aerated, since the fleshy roots of the vanilla have high oxygen requirements. Coniferous bark mixed with live sphagnum moss may also be used.

Watering: The *Vanilla*, like all orchids, need rain-water or treated water. It is very sensitive to overwatering. The container may be allowed to dry out between waterings. The plant will react negatively to cold water poured over its roots by ceasing growth. It survives the cold season without difficulty if its roots are kept almost dry. Spraying with water at room temperature will encourage growth.

The characteristic
fragrance of
vanilla pods
appears only after
fermentation has
been effected.

Feeding: Feed every 4 weeks from March to September. Either use a commercial preparation or a diluted cow-manure. Other fertilizers are not recommended as they will lead to mineral salt accumulations in the soil and the roots will be damaged.

Maturing, harvesting: Well kept *Vanilla* plants will flower in their third year. When the fruit has formed, wait until the pod is slightly yellow at the apex and begins to burst.

You could try to ferment the pods and encourage the formation of *Vanilla* crystals by alternating the application of damp heat with drying out. Unless this is done, the pods will be utterly lacking in taste and aroma. When they are ready for use, they will have shrunk to the width of knitting needles, be dark brown and smell delicious. The vanilla crystals will be readily visible on the outside of the pod.

Propagation, cultivation: Take the terminal shoot of a vanilla plant to a length of about 40 cm. The lower part, roughly half of it, which is to lie in the soil is stripped of its leaves. Lay this cutting in an appropriately large pot so that the covered part is several centimetres deep. The other leafy section is attached to a stick vertically. Cover with a polythene bag, having watered it slightly.

At a temperature of 20 to 25°C in a shady position roots will develop after 3 to 4 weeks. This is indicated by the new growth. Seeds, which are as good as unobtainable, usually give poorish plants.

Ginger
Zingiber officinale
Family: Zingiberaceae

Ginger rhizomes are generally available and can be used for propagation provided they are fresh.

Origin: This was tropical India and Pakistan. Some varieties are grown in South East Asia. Today it is cultivated on a large scale in China, India, Japan, the West Indies and Africa.

Habit: *Zingiber* is a reed-like plant. From a subterranean fleshy knotty rhizome arise leafy shoots up to 100 cm long with alternate leaves 15–20 cm long and about 2 cm broad. The inflorescences are on shorter scaly stems and take the form of dense spikes with yellow-green flowers. Fruits are hardly ever produced.

Habitat: From February to October ginger needs a light, warm place. In the warm season the temperatures may be allowed to rise to 30°C. It may be grown in the living room, but only in a light south or west facing window. In the open air it requires a sunny sheltered spot. It will not tolerate lengthy periods of rain. Since it dies off in late autumn, it can overwinter in a cool room not below 10°C. It remains in

Ginger is a reed-
like plant (right).
The knotty
rhizome (below)
is of great
culinary

importance. It is
used fresh, as a
preserve or –
dried and ground
– as a spice.

the same pot, in the same soil in which it has stood in summer, kept completely dry.

Soil: It will do well in standard potting mixture.

Watering: During the growth period ginger requires an evenly moist soil. When the leaves begin to go yellow in autumn, it must be kept drier. It will not tolerate tap water in the long term. The rhizome must be kept completely dry in winter until ready to grow again. Planted in the greenhouse, ginger will tolerate moist conditions during its resting period: it will put out new growth as ground temperatures rise.

Feeding: Given the standard potting mixture no special feed is required.

Maturing, harvesting: When the leaves have turned yellow in autumn, the rhizome can be lifted. Fresh ginger may be prepared in a sugar solution: there are various recipes. Some people eat it fresh, cut in small slices, with roast meat, ham, etc. After careful cleaning it

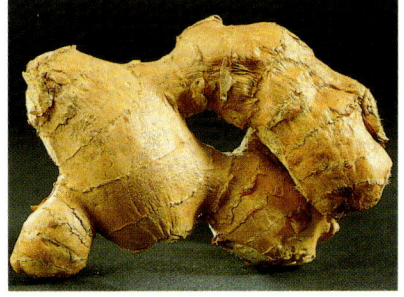

can also be dried, but must not be exposed to sunlight. Drying takes about 14 days and the rhizome loses up to 70% of its weight. It develops a camphor-like smell and a refreshing pepper-like taste.

Propagation, cultivation: The rhizomes which we buy for propagation purposes must be turgid and with no decayed parts. They are laid flat on the soil and only slightly covered. A polythene bag is used to retain the humidity. Water only slightly. Keep in a shady place at c. 20°C.

Depending on the freshness of the rhizomes, root development will begin after 14 days to 6 weeks. When the leaf shoots begin to grow, remove the covering. Rhizomes that have overwintered need to have the old soil removed from them, and are then planted as described above.

Acknowledgements for illustrations

Apel, H., Baden-Baden: pp. 24, 25, 33, 36, 40, 42, 44, 46 (all), 55, 67, 81 (right), 84 (right), 88, 99, 106, 117, 118, 120.

Bärtels, A., Waake: pp. 17, 77, 96, 113 (left).

Bechtel, H., Heimbach: pp. 14 (all), 23, 28 (above), 35, 39, 73, 75 (below), 84 (left), 95, 100, 111, 113 (right).

Espig, G., Stuttgart: p. 57, 60 (right), 75 (above), 92, 101, 109, 115, 123.

Frantz, J., Tübingen: p. 7.

Haase, M., München: p. 60 (left).

Jenuwein, H., Aystetten: pp. 9, 15, 21, 28 (below), 29, 37, 64, 71 (left), 81 (left), 82, 91, 93, 97, 98, 105, 110.

Laux, E. H., Biberach: pp. 38, 63, 71 (right), 76, 107, 114, 121.

Log id, Tübingen pp. 2, 11.

Morell, E., Dreieich: pp. 31, 51.

Reinhard, H., Heiligenkreuz-steinach-Eiterbach: Cover illustration, pp. 19, 102 (left).

Seibold, H., Hannover: pp. 53, 69, 78, 87.

Seidl, S., München: p. 89.

Van Dijk Co., Holland: p. 59.

Walke, M., Karben: p. 102 (right).

Wetterwald, M. F., Offenburg: p. 123 (all).

Index

Page numbers with *
indicate illustrations.